Upgraded Chemical Process Technology and Efficient Equipment

Upgraded Chemical Process Technology and Efficient Equipment

Amalesh Sirkar
Professor
Chemical Engineering Department
Haldia Institute of Technology
Haldia, West Bengal

PUBLISHING FOR ONE WORLD

NEW AGE INTERNATIONAL (P) LIMITED, PUBLISHERS
New Delhi • Bangalore • Chennai • Cochin • Guwahati • Hyderabad
Jalandhar • Kolkata • Lucknow • Mumbai • Ranchi
Visit us at **www.newagepublishers.com**

Published by New Age International (P) Ltd., Publishers
First Edition: 2011

Branches:

- 37/10, 8th Cross (Near Hanuman Temple), Azad Nagar, Chamrajpet, **Bangalore**-560 018. Tel.: (080) 2675 6823, Telefax: 2675 6820, E-mail: bangalore@newagepublishers.com
- 26, Damodaran Street, T. Nagar, **Chennai**-600 017. Tel.: (044) 24353401, Telefax: 24351463 E-mail: chennai@newagepublishers.com
- CC-39/1016, Carrier Station Road, Ernakulam South, **Cochin**-682 016. Tel.: (0484) 2377004, Telefax: 4051303 E-mail: cochin@newagepublishers.com
- Hemsen Complex, Mohd. Shah Road, Paltan Bazar, Near Starline Hotel, **Guwahati**-781 008. Tel.: (0361) 2513881 Telefax: 2543669, E-mail: guwahati@newagepublishers.com
- 105, 1st Floor, Madhiray Kaveri Tower, 3-2-19, Azam Jahi Road, Nimboliadda, **Hyderabad**-500 027. Tel.: (040) 24652456, Telefax: 24652457, E-mail: hyderabad@newagepublishers.com
- RDB Chambers (Formerly Lotus Cinema)106A, 1st Floor, S.N. Banerjee Road, **Kolkata**-700 014. Tel.: (033) 22273773, Telefax: 22275247, E-mail: kolkata@newagepublishers.com
- 16-A, Jopling Road, **Lucknow**-226 001. Tel.: (0522) 2209578, 4045297, Telefax: 2204098 E-mail: lucknow@newagepublishers.com
- 142C, Victor House, Ground Floor, N.M. Joshi Marg, Lower Parel, **Mumbai**-400 013. Tel.: (022) 24927869 Telefax: 24915415, E-mail: mumbai@newagepublishers.com
- 22, Golden House, Daryaganj, **New Delhi**-110 002. Tel.: (011) 23262370, 23262368, Telefax: 43551305 E-mail: sales@newagepublishers.com

---

ISBN : 978-81-224-3034-9

Rs. 395.00

C-10-09-5057

Printed in India at Glorious Printers, Delhi.
Typeset at River Edge, New Delhi.

PUBLISHING FOR ONE WORLD
NEW AGE INTERNATIONAL (P) LIMITED, PUBLISHERS
4835/24, Ansari Road, Daryaganj, New Delhi-110002
Visit us at **www.newagepublishers.com**

Dedicated to

My wife, Bulbul
Daughter, Nabaneeta
Son-in-law, Jayabrata
&
Beloved Grand Daughter Nihashi

Preface

This book aims to encourage creative activity among practicing chemical engineers involved in the industry today. The need for creativity in the upgradation of existing technologies, reduction of equipment size and in the efficiency enhancement of process equipment (through new design features) is justified primarily by the potential for reducing costs and improving product quality. Since these two objectives have long been the cornerstones of the global engineering faculty. I am enthusiastic about the prospect of invoking fresh ideas, innovation and sheer creative energy in the minds of my readers.

The book has been divided into two sections — "Chemical Process Technology and Creativity" and "Equipment Design and Creativity". In the former section traditional/dated process concepts and operating parameters have been presented to begin with. Subsequently, I have dwelled upon newer process concepts and techniques that typically lead to the upgradation of the older technologies. In this manner, I hope to assist the reader develop an exciting insight into the roots of ideas and techniques that have the potential for furthering existing or traditional technologies. The objectives of this book would be considered successfully met if the reader looks to question afresh the "established" protocols of chemical engineering and applies these creative/innovative ideas and techniques at the time of exploration.

The section on Equipment Design and Creativity elaborates on certain key design features that can effectively reduce the size of equipment and, in some cases, improve their efficiency as well. Occasionally mathematical equations have been employed to determine the size of the equipment while elsewhere, the advantages of the new design features have been presented at a conceptual level. I hope that this judicial mixture of technical details and high level concepts will help readers with varying knowledge needs and feel equally comfortable with the contents and the message of this book.

Examples justifying both, improved technology as well as improved equipment design, accompany most of the chapters. Here again, these instances have been hand picked from experimental domains as well as from purely theoretical analysis — thereby covering both thought as well as facts adequately.

Over the many years of research, engineering, consulting and teaching experience that together comprises my professional career, I have come to believe in the power of the creative element. At the same time, I feel that practicing engineers rarely do justice to this virtue. Hence, above all, I aspire that my reader emerges from this academic experience with a newfound interest in creative philosophy as applicable to the engineering domain.

Author

Acknowledgements

I sincerely thank Mr. Somak Jyoti Sahu, Sr. Lecturer of Chemical Engineering Department of Haldia Institute of Technology for contributing to Chapter 1 & 2 of Section II in terms of developing the computer programmes as well as for helping in drawing many figures of different chapters.

My heartfelt thanks go to Mr. Sebak Ranjan Roy (Ex-Assistant Professor of Biotechnology Department of Haldia Institute of Technology) for valuable contribution to Chapter 1 of Section I.

Last but not the least I thank Prof. Asit Kumar Mitra of Chemical Engineering Department of Haldia Institute of Technology (Ex-Professor of Jadavpur University & IIT Kharagpur) for his valuable suggestions after review of some of the important chapters of the book and for suggesting an appropriate title of the book.

Author

Contents

SECTION II EQUIPMENT DESIGN AND CREATIVITY

SECTION I

CHEMICAL PROCESS TECHNOLOGY AND CREATIVITY

Section I

CHAPTER 1

Glucose to Ethanol

The conventional process for manufacture of ethanol from molasses is by fermentation using yeast followed by distillation. This process produces a maximum of 95·6% by ωt ethanol due to azeotrope formation with water in the distillation step. In order to get absolute ethanol from the azeotrope one has to add a 3rd component *i.e.,* benzene to break the azeotrope which is capital intensive and operating cost is high because it requires 3 distillation columns to separate ethanol from water. An alternative and cheaper process has been invented as described below. However, there are other alternatives also as follows:

1.1 ALTERNATIVE WAYS OF GETTING ABSOLUTE ALCOHOL[1]

1. Alcohol can be dried with desicants such as calcium oxide which reacts with water present, forming calcium hydroxide. Other types of desicants are cellulose or corn starch which provide greater energy saving than calcium hydroxide, because less energy is needed to dry cellulose or corn starch than to drive the water off calcium hydroxide. If cracked corn is used as the desicant then it does not even have to be redried, but can be fermented wet to make more sugar and hence more alcohol.
2. Another novel technique for separating alcohol from water solution is use of rayon which retards the movement of water vapour, but allows organic vapours to travel freely. An endless loop of yarn fibres is pulled slowly through a tube into which alcohol — water vapours are introduced. The water is removed by the yarn which is then dried by heating, and pure alcohol vapour is recovered from the other end of the tube.
3. Alcohol can be separated from water with membranes that are comparatively impermeable to alcohol but allow the water to pass through. But, the pressure required is too high to be practical.
4. Zeolites — molecular sieves selectively adsorb molecules smaller than a certain size. These porous rocks might be useful in alcohol recovery. A promising material is clinoptilolite, a naturally occurring zeolite with holes that water but not alcohol, should fit into.

5. A recently developed technology[8] of anhydrous ethanol production is a continuous process of in-situ extraction of ethanol in the fermentation broth through continuous injection of an organic solvent and separation of ethanol from solvent by distillation. Here, inhibition caused by ethanol and other inhibitors is eliminated, causing an increase in ethanol yield.

A New Process

We know that ethanol is soluble in many high molecular weight organic solvents. Therefore, can we not extract ethanol being produced in the aqueous phase into such an organic solvent?

1.2 THE CRITERIA FOR A SUITABLE ORGANIC SOLVENT

1. It has to be immiscible with water.
2. Distribution coefficient of ethanol in the solvent has to be high.
3. Its boiling point should be high such that separation of ethanol is easy.
4. Its density should be higher than ethanol but lower than water so that there will occur density convection in both the phases due to unstable density gradient [2, 4 & 5] and hence rate of mass transfer from bulk to interface and from interface to bulk could be enhanced.
5. Moderately low interfacial tension so that interfacial mass transfer coefficient is high for quick transfer of ethanol from aqueous to organic phase.
6. Low viscosity is preferred so that natural convective mass transfer rate is higher in the organic phase.
7. Toxicity of yeast w.r.t. the solvent is low.

A list of such solvents is given below

1. Ethyl benzene
2. Ethyl oleate
3. Ethyl caprylate
4. *N*-butyl propionate
5. *n*-Decanol.

Extraction Time

Unstable density gradients increase the rate of ethanol transfer from bulk to interface and from interface to bulk in aqueous and organic phases respectively due to convective mass transfer which is higher than diffusional rate of transfer[3-5]. To further reduce the time of extraction of ethanol from aqueous to organic phase a novel technique can be used as follows:

Due to unstable density gradient which occurs in both the phases, the liquid from the interfacial region either sinks in the aqueous phase or rises to the phase boundary in organic phase. To expedite this, liquid may be withdrawn from the interfacial region containing less ethanol and injected to the region below maximum ethanol concentration in the aqueous phase. Similarly, in the organic phase[10] withdraw liquid from interfacial region containing more ethanol and inject above the region of minimum concentration. This will result in reducing ethanol concentration in bulk of aqueous phase and increase ethanol concentration in bulk of organic phase. Simultaneously, withdraw liquid from the region of maximum concentration of ethanol and inject in the vicinity of interface in the aqueous phase and withdraw liquid from the region of minimum concentration of ethanol and inject it in the vicinity of interface on the organic side. The effect is increase in concentration gradient in the vicinity of the interface in both phases, resulting in higher transfer rate to the interface on aqueous side and to the bulk in organic phase.

Mathematically, the concept can be expressed as follows:

1. Aqueous Phase

Increase in concentration of ethanol in aqueous phase = rate of generation of ethanol due to fermentation — Rate of transfer of ethanol from aqueous to organic phase.

Where, Rate of transfer of ethanol from aqueous to organic phase = Rate of decrease of ethanol concentration below the region of maximum ethanol concentration – Rate of increase of ethanol above the region of maximum concentration + Rate of diffusion at the interface

$$\therefore \quad \frac{dC_{al_{av}}}{dt} = 2k_r C_{Glu_i} - k_i\left(C_{al_{b_i}} - \frac{C_{org_i}}{K_d}\right)\frac{A_i}{V_{aq}} \qquad \text{...(1)}$$

$$\frac{A_i}{V_{aq}} k_i \left(C_{al_{b_i}} - \frac{C_{org_i}}{K_d}\right) = \text{Rate of transfer from aqueous to organic phase}$$

$$= \frac{d}{dt}\Bigg[\underbrace{C_{al_{lh}} - \left\{\frac{W_{al_i} C_{al_{b_i}} dt + V_{lh} C_{al_{lh}}}{W_{al_i} dt + V_{lh}}\right\}}_{\text{Rate of decrease of alcohol}}\Bigg] - \frac{d}{dt}\Bigg[\underbrace{C_{al_{up}} + \left\{\frac{V_{up} C_{al_{up}} + W_{al} C_{al_{max}} dt}{V_{up} + W_{al} dt}\right\}}_{\text{Rate of increase of alcohol}}\Bigg]$$

$$+ \underbrace{D_{AB} \frac{\partial^2 C_{al_b}}{\partial x^2}}\Big|_{x=x_i,\, C_{al_b} = C^*_{al_{b_i}}} \qquad \text{...(2)}$$

Rate of transfer at the interface

The third term on the R.H.S. of equation (2) can be evaluated as follows:

Since one has to estimate the concentration gradient at the interface on aqueous side, one has to know the rate of withdrawal of liquid from the region of maximum concentration of ethanol and its injection above this region.

$$\text{Where } C^*_{al_{b_i}} = \frac{V_i C_{al_{b_i}} + \dfrac{W_{al}}{V_{up}} V_i C_{al_{max}} dt}{V_i + \dfrac{W_{al}}{V_{up}} V_i dt} \qquad \text{...(3)}$$

$\therefore\ C^*_{al_{b_i}} > C_{al_{b_i}}$. Hence, the gradient at the interface is greater. So, mass transfer is greater across the interface.

Similar equations for increased rate of mass transfer from interface to bulk of the organic phase can be developed as per the concept described earlier.

Essentially, what one is trying to achieve is what the stirrer does *i.e.,* either transfer of enriched solute from bulk to interface and returning layer of depleted solute from interface to bulk or transfer of depleted solute from bulk to interface and returning layer of enriched solute from interface to bulk. The circulation achieved by stirring is supplemented by withdrawal and injection process.

In the present system the functioning of the yeast is not hampered whereas with stirrer especially at high speed, yeasts' natural functioning may be disturbed due to beating of yeast by the solid impeller surface causing breakage of cell wall. Besides, the disadvantages of the use of stirrer is that it may cause localized emulsification at the interface due to disturbance created by the stirrer.

1.3 THE FERMENTATION TIME CAN BE CUT DOWN BY INNOVATIVE TECHNIQUES AS FOLLOWS

(1) The rate of transfer of glucose from bulk to yeast cell wall and the rate of transfer of alcohol from yeast cell wall to bulk of aqueous phase can be increased by using inert particles of particular shape and vibrating the fermentor in lateral direction. Figures 1.1 and 1.2 show two different shapes. The mechanism of increased rate of mass transfer is as follows:

Figure 1.1

Direction of back and forth movement of fermentor containing the particles

Note: The shaded portions are solid strips only.

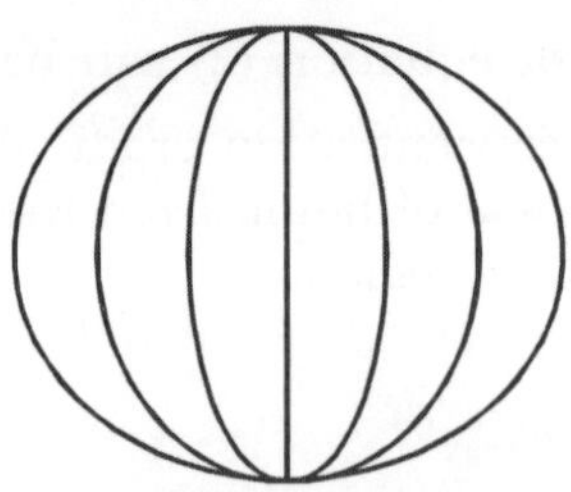

Figure 1.2

Note: The curved lines are actually solid strips.

When the fermentor moves back and forth in the lateral direction the liquid inside the vessel moves through the open space (gap). However, the particle may also move with certain velocity depending on its weight. Since, the velocity with which liquid moves through the gap is higher than with which the particle moves there is a velocity gradient in the unsteady boundary layer which is responsible for increased rates of mass transfer thus, cutting down the fermentation time. The boundary layer thickness is time dependent because motion of particle starts from rest in each direction. Also it is oscillatory flow of liquid within the fermentor.

(2) Use of internal in the fermentor (Fig. 1.3) in the aqueous phase and back and forth movement of the vessel may also increase the rate of fermentation as follows:

The fermentor consists of baffle plates installed vertically (parallel to the wall of fermentor) in the aqueous phase. When the fermentor moves, the liquid inside the vessel moves and creates a velocity gradient in the boundary layer next to the plates. This velocity gradient is responsible for increased rates of mass transfer from bulk to cell wall of glucose and from cell wall to bulk of alcohol thus, cutting down the fermentation time. It is oscillatory flow within the fermentor.

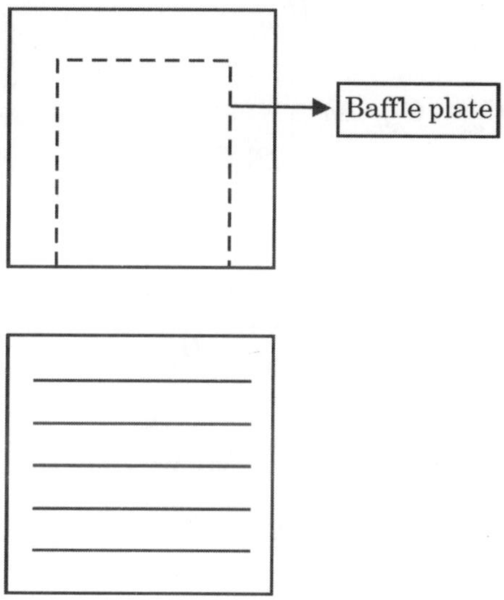

Figure 1.3

Direction of back and forth movement of fermentor

By the process described above one can produce both absolute ethanol (high price product) by distilling the organic phase and an ethanol product by distilling the aqueous phase, the maximum concentration being 95·6% by wt. By varying the type of solvent and its quantity the amount extracted into the organic phase can be controlled. The advantages of the new process are as follows:

1. Fermentation time may be cut down drastically thus, requiring much lower capacity vessels to carry out fermentation[6].
2. Inert particles can be used continuously without any replacement. It is only one time investment. Alternatively, the fermentor can be fabricated with internals which is also one time investment only.
3. Power requirement for the fermentor due to vibratory motion is not much because the vessel is small in size.
4. There is no power requirement for pumping liquid within the fermentor because of the suction head.

5. Substantial amount of ethanol can be extracted into the organic phase.
6. Although the solvent may be expensive there is no loss of it and it can be continuously reused. Negligible loss of it may occur with the product (absolute ethanol). So, basically it is one time investment.

1.4 MATHEMATICAL MODELING OF THE FERMENTATION PROCESS USING A SOLVENT

Rate of increase of ethanol concentration in aqueous phase = Rate of generation of ethanol – Rate of transfer of ethanol into organic phase

$$\therefore \quad \frac{dC_{al_{av}}}{dt} = f(t) = \frac{\sum_{h=0}^{H_{aq}} \Re_{al} A_{yeast}(C_{al_i} - C_{al_b})\Delta h}{H_{aq}} - \Re_i \left(C_{al_{b_i}} - \frac{C_{org_i}}{K_D} \right) \frac{A_i}{V_{aq}} \quad ...(4)$$

where
$$\frac{\sum_{h=0}^{H_{aq}} \Re_{al} A_{yeast}(C_{al_i} - C_{al_b})\Delta h}{H_{aq}} = 2\Re_r C_{Glu_i} \quad ...(5)$$

Also,
$$-\frac{dC_{Glu_{av}}}{dt} = \Re_r C_{Glu_i} = \frac{\sum_{h=0}^{H_{aq}} \Re_{Glu} A_{yeast}(C_{Glu_b} - C_{Glu_i})\Delta h}{H_{aq}} = g(t) \quad ...(6)$$

Furthermore,
$$\frac{dC_{org_{av}}}{dt} = \Re^T_{org}(C_{org_i} - C_{org_{av}})\frac{A_i}{V_{org}}$$

$$= \Re^T_{aq}(C_{al_{av}} - C_{al_{b_i}})\frac{A_i}{V_{aq}}$$

$$= \Re_i \left(C_{al_{b_i}} - \frac{C_{org_i}}{K_D} \right) \frac{A_i}{V_{aq}} \quad ...(7)$$

where $\Re_i = A^{-\sigma_i F/RT}$ (Reference 9)

The following could be determined using the mathematical model after measuring the values of $C_{Glu_{av}}, C_{al_{av}}, C_{org_{av}}, \sigma_i, C_{al_{b_i}}, C_{org_i}$ at different time of fermentation. Also, C_{Glu_b} and C_{al_b} are to be measured at different height of aqueous phase and at different time of fermentation.

1. $\Re_r$ as a function of time
2. $\Re^T_{aq}$ and $\Re^T_{org}$ in the form of dimensionless groups *i.e.*, Sherwood No. as a function of Schmidt No. and Rayleigh No.
3. C_{al_i} and C_{Glu_i} as function of time

In the mathematical model the following assumptions could be made for evaluation of $\Re_r, \Re^T_{aq}, \Re^T_{org}, C_{al_i}$ and C_{Glu_i} :

1. C_{al_i} and C_{Glu_i} are not changing appreciably w.r.t. height of the aqueous phase. This assumption makes mathematical analysis simple.
2. A_{yeast} is a function of time of fermentation because of cell growth and decay.
3. $\Re_{al}$ and $\Re_{Glu}$ could be evaluated from the following relationship assuming stagnant system although evolution of CO_2 may create slight turbulence:

   $$\frac{\Re_{al} d_{yeast}}{D_{AB}} = 2.0 \text{ and } \frac{\Re_{Glu} d_{yeast}}{D_M} = 2.0$$

   where d_{yeast} is floc size which can be determined from floc size vs. rpm correlation(7) extrapolated to 0 rpm.
4. It may be assumed that yeast flocs are uniformly distributed in the aqueous phase although there is no stirring. There may be a tendency for the flocs to settle near the bottom of the fermentor but, due to evolution of CO_2 they are probably pushed upwards and kept suspended except near the end of fermentation when CO_2 evolution rate decreases.

   In the above mathematical model withdrawal and injection scheme as elaborated by eqns. (1), (2) and (3) in aqueous phase and similar equations for organic phase may be incorporated if experiment to cut down extraction time is carried out during fermentation.

Experimental Verification of the Process

Some preliminary experiments were carried out using two solvents namely ethyl benzene and ethyl oleate. It has been found that appreciable amount of alcohol was present in the organic phase at the end of fermentation. However, conversion of glucose was low w.r.t. ethyl benzene compared to ethyl oleate, mainly because of higher level of toxicity of ethyl benzene which retarded fermentation.

NOMENCLATURE

A	–	Constant
A_i	–	Area of interface formed by two phase
A_{yeast}	–	Active surface area of yeast per unit volume of fermentation broth
$C_{Glu_{av}}$	–	Average concentration of glucose in fermentation broth
C_{Glu_b}	–	Concentration of glucose in the bulk of fermentation broth
C_{al_b}	–	Concentration of alcohol in the bulk of fermentation broth
C_{Glu_i}	–	Concentration of glucose at the surface of yeast
$C_{al_{av}}$	–	Average concentration of alcohol in fermentation broth

C_{al_i}	–	Concentration of alcohol at yeast surface
$C_{al_{b_i}}$	–	Concentration of alcohol at the interface on aqueous side
$C_{al_{up}}$	–	Average concentration of ethanol in the region of aqueous phase above maximum alcohol concentration
$C_{al_{lh}}$	–	Average concentration of alcohol in the region of aqueous phase below maximum alcohol concentration
$C_{al_{max}}$	–	Point of maximum concentration of alcohol in the aqueous phase
d_{yeast}	–	diameter of yeast floc
D_M	–	Molecular diffusivity of Glucose in aqueous phase
D_{AB}	–	Molecular diffusivity of alcohol in aqueous phase
F	–	Constant
h	–	any height of aqueous phase from bottom of fermentor
H_{aq}	–	Height of aqueous phase
$\Re_{al}$	–	Mass transfer coefficient of alcohol from yeast surface to bulk
$\Re_{Glu}$	–	Mass transfer coefficient of glucose from bulk to yeast surface
K_D	–	Distribution coefficient to alcohol between aqueous and organic phase
$\Re_r$	–	Rate constant for conversion of glucose to alcohol
$\Re_i$	–	Interfacial mass transfer coefficient from aqueous to organic side
$\Re^T_{aq}, \Re^T_{org}$	–	Convective mass transfer coefficient of alcohol in aqueous and organic phase respectively
t	–	Time
T	–	Absolute temperature
V_{aq}, V_{org}	–	Aqueous and Organic phase volume respectively
V_{lh}	–	Volume of aqueous phase below the region of maximum alcohol concentration
V_{up}	–	Volume of aqueous phase above the region of maximum alcohol concentration
V_i	–	Volume of aqueous phase near the interface
W_{al}	–	Rate of withdrawal of liquid from the point of maximum alcohol concentration in aqueous phase
W_{al_i}	–	Rate of withdrawal of liquid from the interface on aqueous side

Greek letters: σ_i – Interfacial tension between the aqueous and organic phase.

REFERENCES

1. Science, Vol. 206, 5th October, 1979.
2. Berg and Morig, *Chem. Eng. Science*, Vol. 24,1969, pp. 937–946.
3. Berg and Haselburger, *Chemical Engg. Science*, Vol. 26, 1971, pp. 481–485.
4. Sirkar, A. and Walas, S.M. *Chem. Engg. Sci*, Vol. 34, (1979), pp. 741–742.
5. Paruya, S., Manna, T.K. and Sirkar, A. Paper presented at CHEMCON 2003 (December).
6. Indian Patent No. : 173212 by Dr. Amalesh Sirkar.
7. Brendan Brohan and Aiden J. Mclonghlin, *Applied Microbiology and Biotechnology* 20 : (1984) pp. 16–22.
8. Huang, H.J., *Separation and Purification Technology*, 62 (2008), pp. 1–21.
9. Abramzon, A.A., Ostrovskii, M.V. and Frumin, G.T., Zhurnal Prikladnoi Khimii, 41 (1968), pp. 803–810.
10. Indian Patent 201474 (2007) by Dr. Amalesh Sirkar.

Section I

CHAPTER 2

Glucose to Sorbitol

In the conventional process for manufacture of Sorbitol from Glucose, which is a hydrogenation process using Nickel Catalyst, the hydrogen pressure used is 1100 psi. The reason for such a high pressure is low solubility of hydrogen in 40% glucose solution which is normally the feed composition used in a trickle bed reactor. High operating pressure makes the plant expensive: the reactor is expensive, high-pressure pump for glucose solution, high pressure H_2 compressor and the gas/liquid separation section is also expensive because pressure is high for the vessels, exchangers and coolers etc. Therefore, to cut down the cost of the plant, operating pressure should be reduced. The question is how?

2.1 NEW PROCESS DETAILS

We know that there are organic solvents in which H_2 has high solubility compared to H_2O *e.g.* methanol, ethylene glycol etc. Note that these liquids are miscible with water in all proportions. Therefore, we can mix the feed with desired amount of such organic liquid to achieve same solubility of H_2 as in the feed without solvent, thus lowering the operating pressure. In choosing a particular solvent the following properties are to be considered:

1. The liquid should not take part in the reaction
2. The liquid should be easily separable from the reaction products
3. The molecular diffusivity of H_2 in the liquid should be high
4. The molecular diffusivity of glucose in solution should be high
5. Viscosity of the liquid should be low
6. Surface tension of the liquid should be low

Objectives (4) and (5) are obvious from the following correlation for a cocurrent trickle bed reactor[(1)]

$$\ln C_{in}/C_{out} = \eta \Re_c L^{0.32} (SV)^{-0.68} d_p^{0.18} (\mu_L/\rho_L)^{-0.05} (\sigma_C/\sigma)^{.21} \qquad ...(1)$$

where η is catalyst effectiveness factor, which is given by

$$\eta = 3 / \phi_s \ (1/ \tanh \ \phi_s - 1/\phi_s) \quad \text{...(2)}$$

$$\varphi_s = R_P \ (\rho_p \hat{\Re}_c / D^{\text{eff}}_{\text{Glucose}} - H_2O) \quad \text{...(3)}$$

After selecting a liquid the obvious question arises as to how much solvent should be added. For a given capacity plant higher the percentage of the solvent higher is its contribution towards the average solubility of hydrogen in the mixture. Since the product sorbitol is normally 70% in water, if we start with 40% glucose solution in water as mentioned earlier lot of water is to be evaporated after reaction. Therefore, substitute portion of water in the feed by the liquid selected *e.g.* methanol, which has all the 6 characteristics mentioned above. An optimum amount of solvent is to be used on the basis of economic consideration as follows:

1. Higher solvent amount reduces the operating pressure thus, cutting down the cost of reactor, some process vessels, H_2 compressor and glucose feed pump.
2. Higher solvent amount increases the size of the reactor because feed volumetric flow rate increases thus increasing the space velocity.
3. Higher solvent amount increases the diameter of distillation column for separation of alcohol from sorbitol.
4. Higher solvent quantity means higher partial pressure of solvent vapour thus, total pressure increases.

A block flow diagram based on conventional technology and another based on methanol as the solvent for manufacture of sorbitol from glucose are shown in Fig. 2.2 & Fig 2.3.

So far the discussion has been restricted to reduction in operating pressure. Another way to cut down the cost of the plant is to reduce the volume of the reactor. This can be done either by selecting a better catalyst or by some means increase the overall rate of reaction if the same catalyst is used. Here we need a breakthrough in our thinking process.

2.2 REACTOR AND CATALYST SHAPE

The first thing which comes to mind is the use of a fluidized bed reactor where one can reduce the resistance to mass transfer to and form the catalyst by using an ebullated bed reactor where H_2 and liquid move cocurrently up through the bed of catalyst which is under suspension. The resistance to mass transfer is reduced by higher gas and liquid velocity passed the catalyst particle. Here lies the disadvantage of the technique *i.e.* higher throughput of gas, large recycle rate of liquid and higher pressure drop. So, what is the alternative?

Instead of increasing the throughput of gas and liquid to increase the mass transfer rate can we not make the catalyst move so that the adjacent liquid moves! The solution is vibrate the reactor. Here begins the research. It has been found through research that if we use particular shape of the catalyst the overall reaction rate increases substantially. So, here is the breakthrough.

Consider the Nickel catalyst shape like a sphere made up of cylindrical rods with space between two rods as shown below:

Figure 2.1

Mechanism of Overall Reaction

As the catalyst moves due to vibration of the reactor the liquid displaced by the rods rushes through the opening to occupy the space previously occupied by the particle. Liquid displacement causes the following:

1. Due to motion of the liquid past the rod sorbitol produced at the surface of the catalyst will be swept away by the displaced liquid which is faster than molecular diffusion.
2. The reaction of hydrogen with glucose at the catalyst surface occurs at a rate which is dependent on the solubility of hydrogen in the liquid at half of the summation of bulk hydrogen and interfacial hydrogen concentration in the gas film which is higher than conventional process. This is because due to vibration catalyst displaces gas (apart from liquid) in the form of wave through the opening between two rods and there occurs collision between the adjacent waves causing mixing within the gas film thus destroying concentration gradient in the film.
3. The movement of liquid dissipates the heat of reaction from catalyst surface thus preventing any hot spot in the reactor unlike fixed bed. An example of the upgraded technology considering the points discussed earlier is given below.

2.3 CONVENTIONAL PROCESS AND IMPROVED PROCESS

In the conventional technology a 40% solution of Glucose in water is pumped upto about 1100 psi, preheated to 293°F and fed to a trickle bed reactor along with compressed hydrogen. The reactor is filled up with Nickel catalyst. The reaction is a one step hydrogenation process producing sorbitol. Space velocity used is about 1 cc/(hr) (cc catalyst) for nearly 100% conversion. The products from the reactor go to gas-liquid separator section followed by evaporation of liquid stream to produce 70% sorbitol product.

In the upgraded technology[2] we reduce the operating pressure by mixing the feed with liquids such as methanol, ethylene glycol in which hydrogen solubility is higher than in water as discussed earlier.

Since, in the conventional process we have to evaporate the water after reaction, we can start with only that much water, which will finally be with the product, and to reduce the viscosity of this feed and lowering of the operation pressure of the reactor, mix it with either methanol or ethylene glycol. In methanol solubility of H_2 is higher than in ethylene glycol but with methanol the risk of fire hazard is more than in the case of ethylene glycol. But, again it is cheaper to separate methanol than ethylene glycol from reaction products. With methanol it has been mentioned earlier that economic considerations will dictate the optimum amount of it to be added to feed. But one has to make sure that with increase in methanol addition conversion does not decrease from 100% due to increase in space velocity provided the other parameters in equation (1) compensate this increase by required amount.

Now, the objective as outlined earlier goes one step further where catalyst volume is to be reduced by vibrating the reactor using a particular shape of the nickel catalyst mentioned earlier because this technique increases the overall rate of reaction by reducing the diffusional resistances.

So, we have fulfilled two objectives of lowering the operating pressure and reduction of the reactor volume as outlined earlier, which leads to upgradation of the conventional technology for manufacture of sorbitol from glucose.

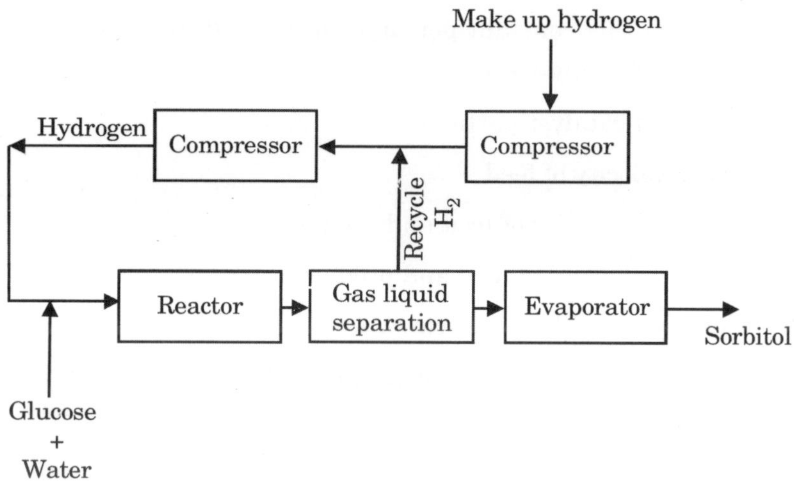

Figure 2.2: Conventional process of sorbitol production

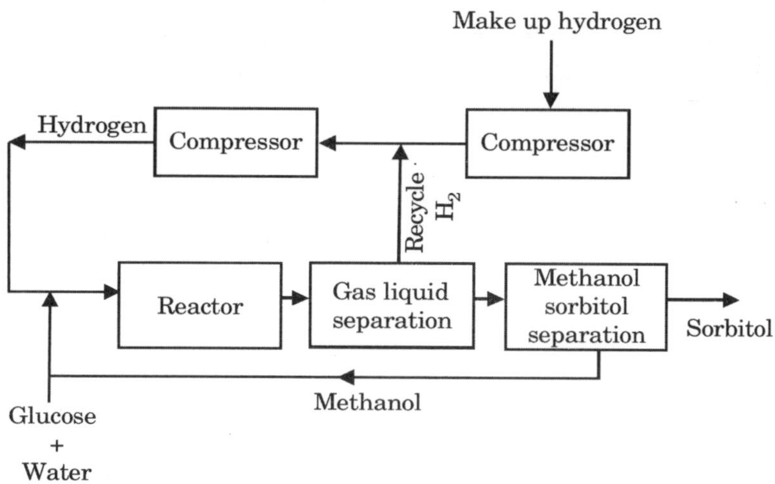

Figure 2.3: Improved process of sorbitol production

NOTATION

C_{in} – concentration of liquid reactant in feed

C_{out} – concentration of liquid reactant in the product

d_p – size of catalyst particle

$D^{eff}_{Glucose\text{-}H_2O}$ – effective diffusivity of glucose in interior of catalyst particle

$\Re_c$ – reaction rate constant

$\hat{\Re}_c$ – reaction rate constant per unit mass of catalyst

L – height of catalyst bed

R_p – radius of catalyst particle

SV – space velocity of feed

μ_L – average viscosity of feed and product

ρ_L – average density of feed and product

ρ_p – density of catalyst particle

σ_C – critical surface tension for nickel

σ – average surface tension of feed and product

REFERENCES

1. Chemical Reactor Design for process plants by Howard F. Rase (Vol. 1), p. 709 (1977).
2. Indian Patent 177026 by Dr. Amalesh Sirkar (1997).

Section I

CHAPTER 3

Polyols from Sorbitol

Hydrogenolysis of Sorbitol[1,2] in presence of Nickel catalyst gives a host of chemicals known as polyols *i.e.* polyhydric alcohols. Out of the host of chemicals a major fraction however consists of glycerol and glycols (both ethylene and propylene glycol).

3.1 PROCESS DESCRIPTION

In the conventional process which is based on bench scale experiment, a 25% (by weight) solution of sorbitol in water obtained directly by hydrogenation of glucose is mixed with 1% lime and then pumped upto 2350 psi and preheated to 464°F before it is fed to a tubular reactor (packed with Raney Nickel catalyst) from top along with hydrogen. It is a trickle bed reactor mode of operation. The space velocity of the feed normally used is 2.3 cc/hr/cc catalyst and the catalyst used is in the form of pellets $\left(\frac{3}{16}'' \, dia \times \frac{1}{8}'' \, ht\right)$. In the bench scale process a reactor dia of 1″ ID and 17″ long was used. In the commercial scale a number of 1″ dia reactors can be installed inside one big reactor cross section. The products consist of 0.7 lb of glycerol and 0.18 lb of glycol per lb of sorbitol fed to the reactor. The balance is other alditols. Figure 3.1 gives a process flow diagram for obtaining various products using conventional separation techniques. In Figure 3.1 we have shown a purge stream and two recycle streams in order to prevent build up of aldoses in the system. Note that aldose will be converted to alditols in Reactor I, but again part of aditols gets converted to aldoses in Reactor II.

Products from hydrogenolysis of sorbitol

Sl. No.	Description	Molecular wt.
1.	Methanol	32
2.	Ethanol	46

Sl. No.	Description	Molecular wt.
3.	1 – propanol	60
4.	2 – propanol	60
5.	Butanol – 1	74
6.	Butanol – 2	74
7.	Water	18
8.	2-3 Butanediol	90
9.	Propylene glycol	76
10.	1-2 Butanediol	90
11.	Ethylene glycol	62
12.	1-3 Butanediol	90
13.	1-4 Butanediol	90
14.	Diethylene glycol	106
15.	Glycerol	92
16.	Pentanediol	104
17.	1-2-4 Butanetriol	106
18.	Triethylene glucol	150
19.	Diglycerol	150
20.	Triglycerol	224
21.	Erythrose	120
22.	Erythritol	122
23.	Arabinose and xylose	150
24.	Arabitol and xylitol	152
25.	Mannose	180
26.	Mannitol	182
27.	Glucose	180
28.	Sorbitol	182

In the conventional process operating pressure is quite high. How can we reduce the pressure? We know that sorbitol is soluble in methanol and that hydrogen has a higher solubility in methanol than water. Therefore, part of the water can be replaced by methanol in the feed to the reactor. This way operating pressure can be reduced. The technique of replacing part of the feed in water by methanol is similar to Glucose to Sorbitol process described in chapter 2.

Similarly, the volume of the reactor can be reduced by the technique described in glucose to sorbitol process.

Methanol based process consists in separation of mixed alcohols and recycling desired amount of methanol to glucose feed to reactor I.

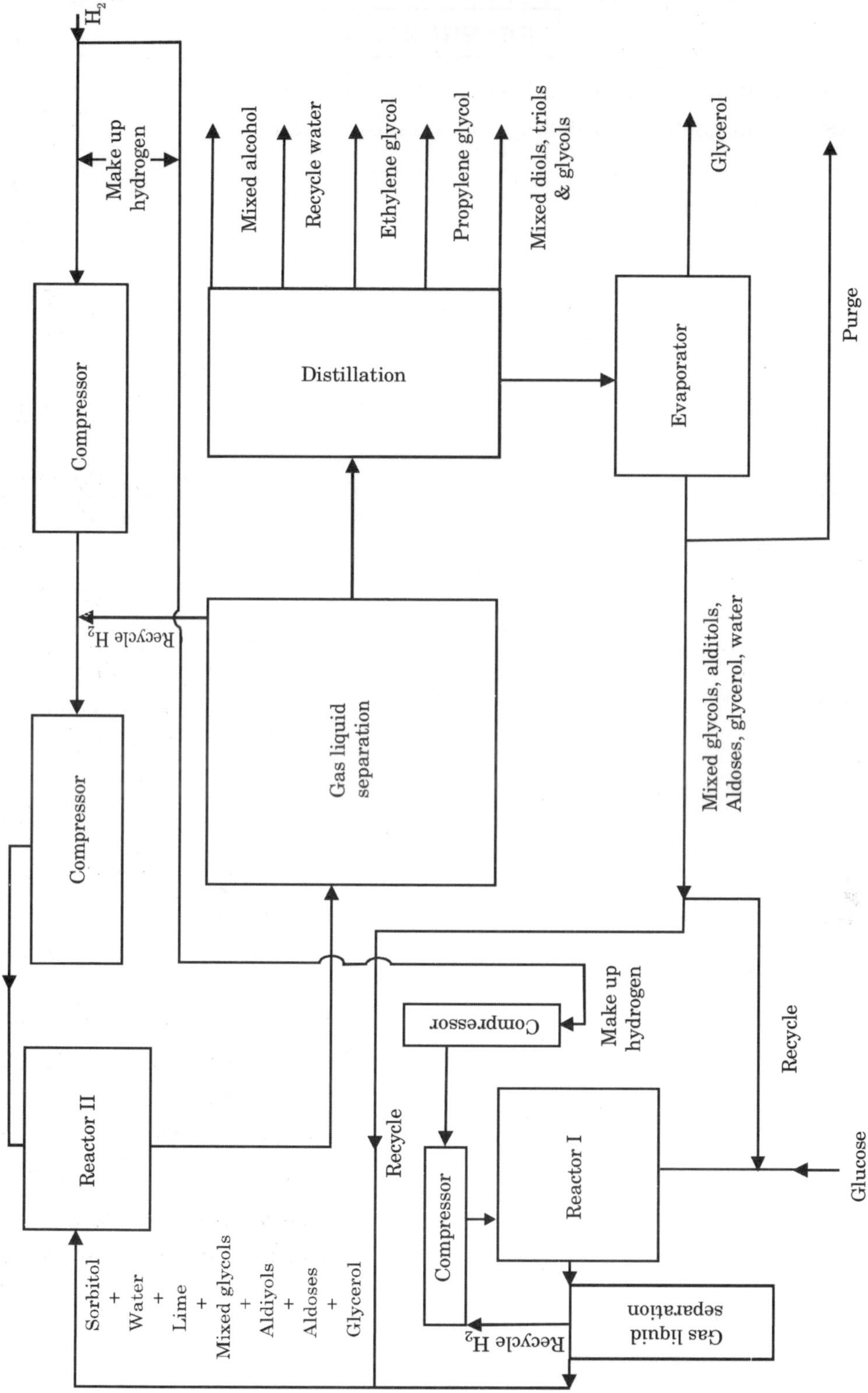

Figure 3.1: Conventional process of polyols production

REFERENCES

1. US Patent 4,338,472 (1982) by Dr. Amalesh Sirkar.
2. US Patent 4,380,678 (1983) by Dr. Amalesh Sirkar.

Section I

CHAPTER 4

Coal Liquefaction

As the price of oil continues to rise and supplies become less certain, countries with little oil reserve *e.g.,* India are looking for ways of producing their energy requirements and petrochemical feedstock for both economic and political reasons. Compared to the coal liquefaction processes developed in the seventies the latest processes based on dry hydrogenation of coal appear to be more competitive with respect to processing of petroleum products because of claims of very high percentage of liquid products[2, 3 & 4].

Coal liquefaction is still virtually untried commercially[1]. The only full-scale plant in operation is the South African Coal Oil and Gas Corp's Sasol I, which has been running since 1956 in Sasolburg. This has been followed by Sasol II and Sasol III both at Secunda. Sasol's Synthol process, a modified Fischer-Tropsch synthesis process involves passing CO and H_2, in a ratio of 1 : 3 over a fluidized bed of powdered iron catalyst. But this process produces only 1.5 bbl of oil per ton of coal.

The other processes developed in the seventies are based on pyrolysis, solvent extraction and catalytic hydrogenation. Amongst these one European process is in commercial operation. A 1600 ton/day lignite plant has been running in Yugoslovia since 1963. Although this liquefaction process is based on pyrolysis method, it in fact produces more gas than liquid. Crushed coal is heated to 450°–600°C and the resultant gases condensed. The product slate is 50% char, 32% gases and 18% liquids. Of the processes nearing commercial utilization most originate from the U.S. Two demonstration plants have been built. On the other hand most European processes are still in laboratory or pilot plant stage.

4.1 VARIOUS COAL LIQUEFACTION PROCESSES DEVELOPED IN USA[1]

Coal liquefaction technologies can be broadly classified into two categories; one stage and two stages methods. The first is typified by the Gulf SRC–II (solvent refined coal) and H – coal processes while an example of the latter is the SRC–I process combined with the cities service/C–E Lummus LC–Fining catalytic hydrocracking. Two stage processes allow for greater flexibility of both product slate and product quality. In addition, two stage processes can be used with a wider variety of coals, enabling them to serve a large potential market.

SRC-I Process (Fig. 4.1)

In this process coal is dissolved in a process-derived solvent and hydrogenated at high temperature and pressure. Sulfur compounds and light hydrocarbons are released as gasses during the dissolving stage. Undissolved coal is then removed and the liquid stream distilled to recover the solvent. This leaves two liquid streams, one heavy and one light.

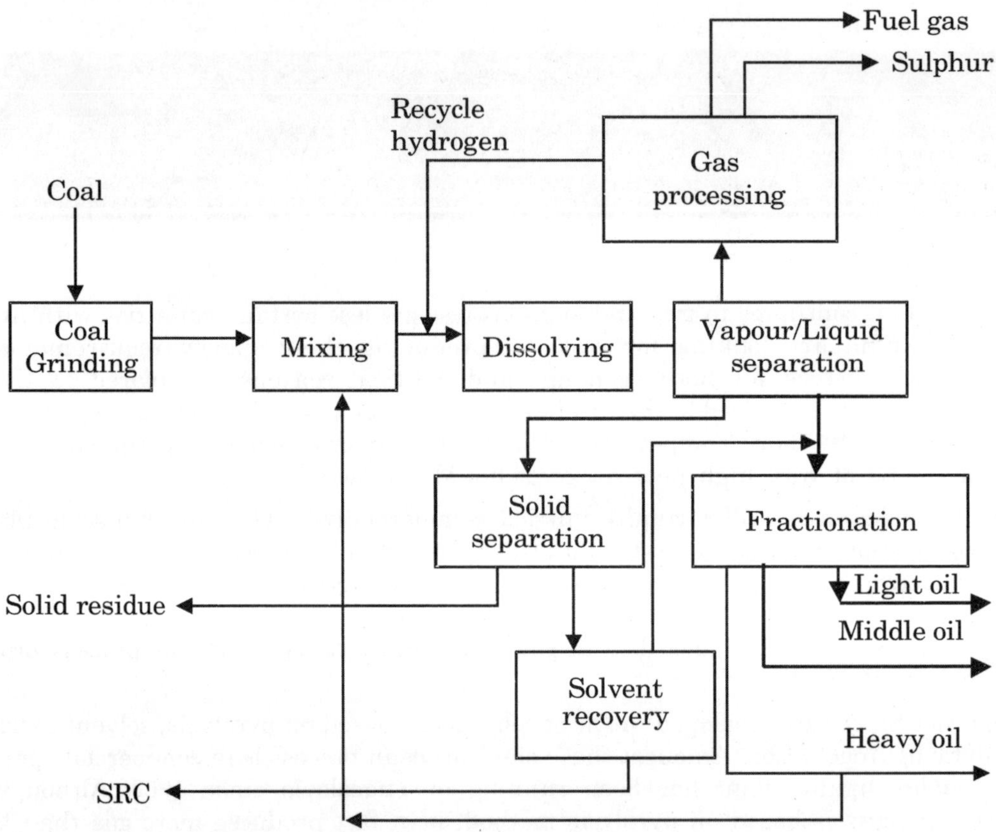

Figure 4.1

SRCñII Process (Fig. 4.2)

This process is similar to SRC–I, but in this case the majority of the product is liquid. This is achieved by recycling part of the undissolved mineral residue which allows more extensive hydrocracking of the dissolved coal to liquid and hydrocarbon gasses. In addition, the process has a number of other special features. These include the gasification of some of the undissolved mineral residue and a portion of the unconverted dissolved coal to produce hydrogen for the process and burning any excess synthesis gas as fuel. In addition, the process produces a substantial quantity of pipe line gas as well as the liquid product. Use of the undissolved residue for hydrogen production, according to Gulf, facilitates the providing of sufficient hydrogen to make significant amounts of C_1–C_4 hydrocarbons while still leaving enough synthesis gas to be used as fuel.

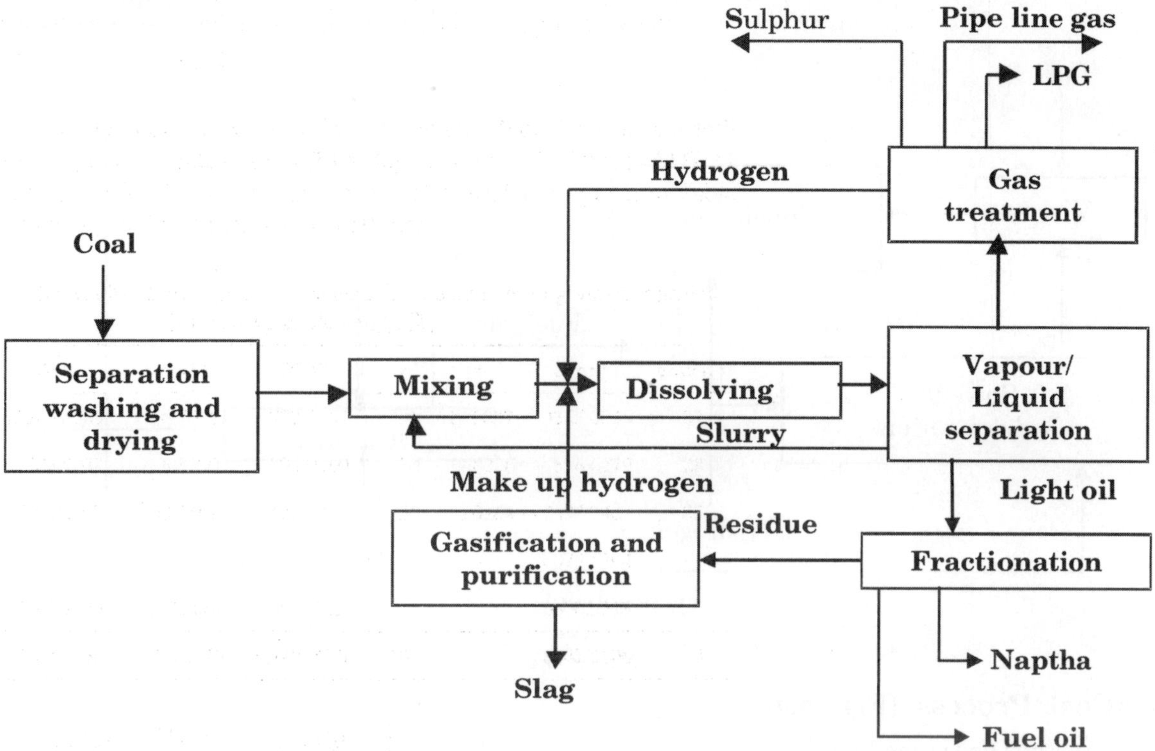

Figure 4.2

Exxon Donor Solvent Process (Fig. 4.3)

In this process coal is first ground and then mixed with the recycle solvent. The mixture is then heated and fed to a reactor where it is liquefied with gaseous hydrogen at around 400°C. Hydrogenation takes place, using hydrogen from both the gas and the solvent. The liquid product is then distilled to separate the recycle solvent and desired product, leaving a residue of undistillable tar, unreacted coal and ash. This is then fed to a Flexicoking unit, where additional liquids and a low value fuel gas are made.

The solvent is recycled, having then rehydrogenated using conventional petroleum hydrotreating technology.

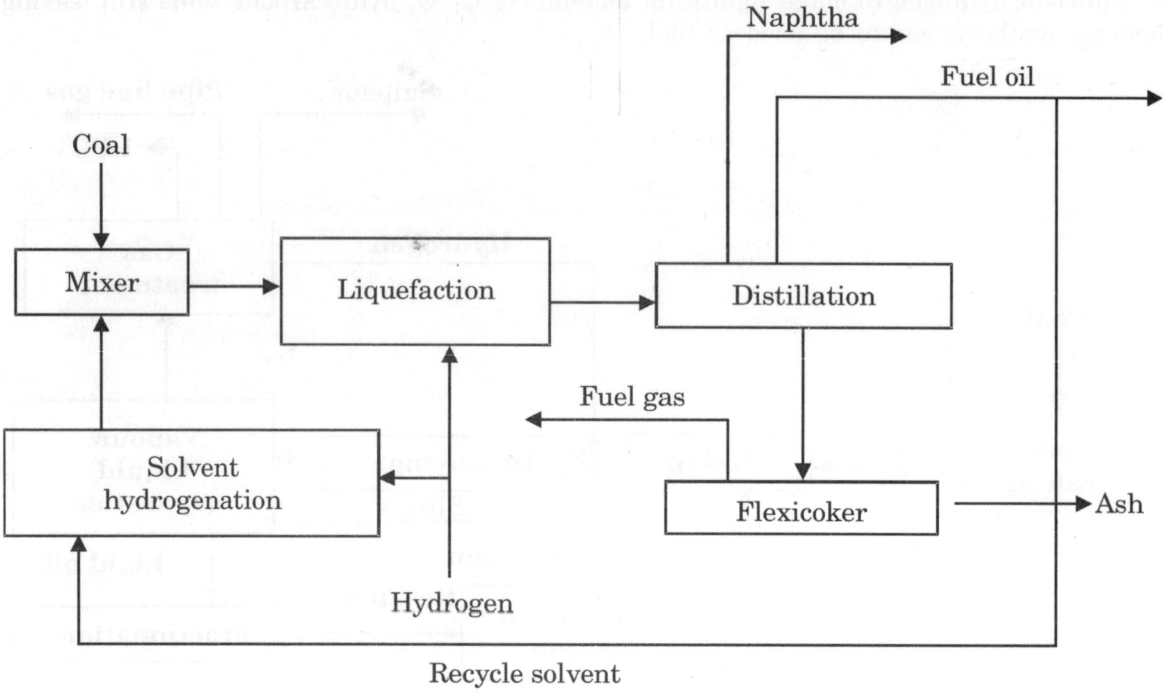

Figure 4.3

HñCoal Process (Fig. 4.4)

Another one stage liquefaction process is Hydrocarbon Research's H–coal process. Like the SRC–II process this involves hydrogenation in a fluidized bed of catalyst after the coal has been ground and slurried. The plant can be operated in one of the two ways, either in the syncrude mode in which case all the products are distillate and can be refined to give gasoline, petrochemicals and light fuel oils, or the fuel oil mode, in which case distillate production is very low and heavy fuel oils predominate. The two differ in the method of hydrogen production. In the former case the hydrogen is produced by gasifying the heavy oil product whereas in the fuel oil mode some unreacted coal is gasified.

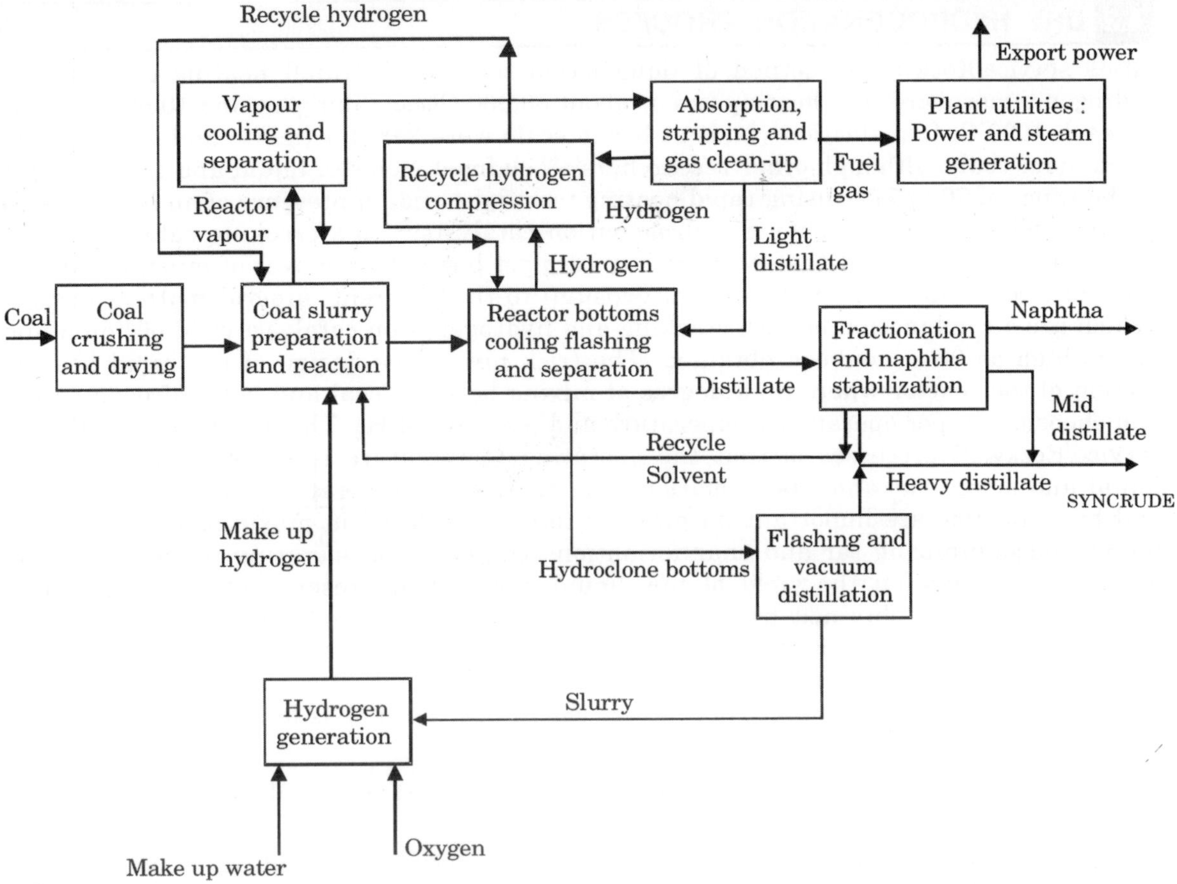

Figure 4.4

4.2 AN IMPROVED LIQUID PHASE ONE STAGE LIQUEFACTION PROCESS (MODIFIED H–COAL PROCESS)

The disadvantage of liquid phase hydrogenation process is that it is limited by the solubility of H_2 in the liquid compared to dry hydrogenation where coal is in direct contact with hydrogen.

How can we increase the solubility of H_2 in liquid? It has been found that if portion of liquid in which coal is slurried, is substituted by lower boiling liquid such as B-T-X in which H_2 solubility is higher, can improve the process by lowering the operating pressure. Lower operating pressure means pumping cost is less and reactor and downstream gas-liquid separators are less costly. Optimum amount of B-T-X will have to be determined on the basis that while increased B-T-X increases H_2 solubility meaning lower H_2 pressure, B-T-X exerts its own vapour pressure thus increasing the total operating pressure. Thus, optimum amount of B-T-X means operating at an optimum total pressure.

Since B-T-X is one of the products of coal liquefaction, a recycle around the reactor can be built up for slurring with coal. Also, at the operating temperature (450°C) B-T-X is fairly stable in the reactor.

4.3 DRY HYDROGENATION PROCESS

Cities service/Rockwell's method of liquefaction involves in rapid heating of coal in a hydrogen atmosphere. The heating rate is about 50,000˚C/sec, taking around 10 milliseconds to reach 1000˚C. Liquefaction yields at bench scale were 24% for lignite and 16% for sub-bituminous coals, although other researchers[2,3] claimed yields of liquid and tar combined in the range of 70 to 77%, using rapid heating technique and, in presence of an impregnated catalyst, which may be either a hydrogenation catalysts[2] or a cracking catalyst[3]. One researcher[4] has claimed that without any catalyst but with removal of mineral matters from raw coal which are poisons (*e.g.*, gypsum) to the inherent mineral matters in coal, (which are a combination of both cracking and hydrogenation catalyst) yield of liquid and tar as high as 98% has been obtained. The trick lies in the design of the rapid heating section of the reactor where high degree of mixing between coal and hot hydrogen occurs and choosing proper operating temperature and pressure of H_2. The disadvantage of cities service/Rockwell processes is that the operating temperature is too high, causing less liquid and more char and gas formation. The operating temperature, operating pressure and heat up time are important parameters to be considered in maximizing the yield of liquid and minimizing tar and char formation. The effect of operating temperature and operating pressure on the yield of tar, liquid and gas in presence of an impregnated catalyst (MoO_3) as obtained by Schroeder[2] is given in Table 1. However, these yield distributions will be different if we use a different reactor where heat up time is less. Schroeder used an entrained downflow reactor heated externally and electrically where heating was not rapid. Examples of reactors where heat up time is less are described later in this chapter.

The importance of quick heating and uniform mixing is mainly due to two reasons as follows:

1. The fact that coal passes through the plastic region[7] at around 325 to 350˚C, and if this region is not passed quickly then the plastic coal (a mixture of non-newtonian liquid and gas) will stick to solid surface in contact with it and jeoparadise the liquefaction process, the reason being that heat transfer to non-newtonian liquid would be poor requiring longer to pass this region.
2. Apart from plastic behaviour of coal another reason for rapid heating is that if time period is long then coal will begin to get converted to tar, liquid and gas during heating. In such a case, the yield of liquid will be reduced due to longer hold-up time which is the total of heating time and reaction time at the operating temperature. The reason for reduction of liquid yield is because longer it stays in the reactor it gets gasified.

It may be noted that although average contact time between hot hydrogen and coal may be say .001 second, meaning rapid heating, depending on the mixing technique product distribution will vary due to variation in the fractions of coal staying longer or shorter in the preheater. The different mixing techniques are elaborated later in this chapter and in a paper[6] by Sirkar and others.

The mechanism of liquefaction and how it affects the product distribution has been elaborated in a paper by Sirkar[5]. The reactions are given below:

Table 4.1: Data on conversion of coal to tar and liquid at different temperature and H_2 pressure, Coal particle size : 70% ñ 200 Mesh and Coal feed rate: 28 lbs/hr./ft^2 [7]

Total H_2 pressure (psi)	°C	Fractional conversion	
		Tar (Max^m)	Liquid (Max^m)
2000	500	0.41	0.29
2000	600	0.16	0.14
4000	500	0.37	0.26
4000	600	0.28	0.25
6000	500	0.38	0.21
6000	600	0.16	0.14

Reactions in solid phase hydrogenation of coal

1. Coal $\xrightarrow[\text{Cracking catalyst inherent in coal}]{\text{Heat}}$ $B_1^\bullet + B_2^\bullet + B_3^\bullet$ + G (gas)
   ($B_1^\bullet$, $B_2^\bullet$, $B_3^\bullet$: Free radicals)
2. $B_1.$ $\xrightarrow[\text{Hydrogenation catalyst inherent in coal}]{H_2}$ T (Tar)
3. $B_2.$ $\xrightarrow[\text{Hydrogenation catalyst inherent in coal}]{H_2}$ L (Liquid)
4. $B_3.$ $\xrightarrow[\text{Hydrogenation catalyst inherent in coal}]{H_2}$ G (gas)
5. T (Tar) $\xrightarrow[\text{Hydrogenation catalyst inherent in coal}]{H_2}$ L (Liquid) + G (gas)
6. L (Liquid) $\xrightarrow[\text{Hydrogenation catalyst inherent in coal}]{H_2}$ G (gas)
7. $B_1. + B_1.$ $\longrightarrow$ C (char)
8. $B_2. + B_2.$ $\longrightarrow$ T (Tar)
9. $B_3. + B_3.$ $\longrightarrow$ G (gas)

4.4 TYPE OF REACTOR FOR LIQUEFACTION OF COAL (FIGS. 4.5 AND 4.6)

There are mainly two types of reactors for liquefaction of coal based on the principle of rapid heating caused by quick mixing of coal and hot hydrogen as follows:

1. Mixing caused by hot hydrogen jet turbulence where the jet entrains coal particles along the expanded jet travel path.

2. Mixing caused by boundary layer separation when a hot hydrogen jet flows over a curved surface.

Based on the above principles of quick mixing various modifications of reactors can be designed. Here we shall describe one example of each principle. For details of modified designs of type (1) please refer to paper by Sirkar and others[6] and see Chapter 2 of Section II.

1. To take advantage of free jet turbulence induced mixing of coal and H_2 we can choose a rectangular box and introduce hot hydrogen jet at left hand corner near the top and coal (pressurized) is fed from top near where hot hydrogen enters. The jet along with coal hits the other side of the rectangular box; bounces backward and downward and like this a jig jag flow of coal and H_2 takes place and coal gets heated quickly by mixing caused by free jet turbulence (Fig. 4.5).

2. Here we take advantage of both free jet turbulence and additional mixing caused by boundary layer separation and eddy formation in the backward direction just behind the spheres. The spheres are fastened to the walls of the square cross-section preheating section (Fig. 4.6).

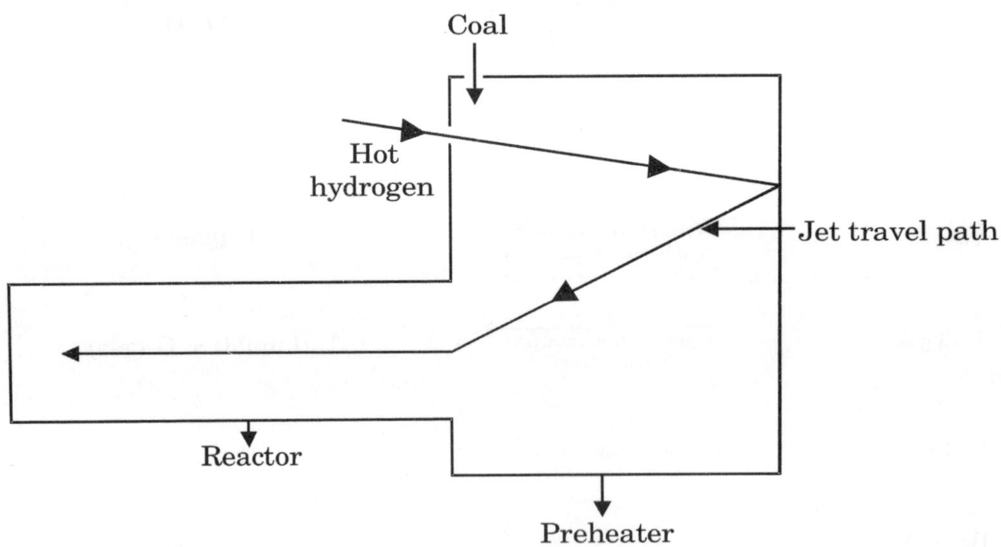

Figure 4.5

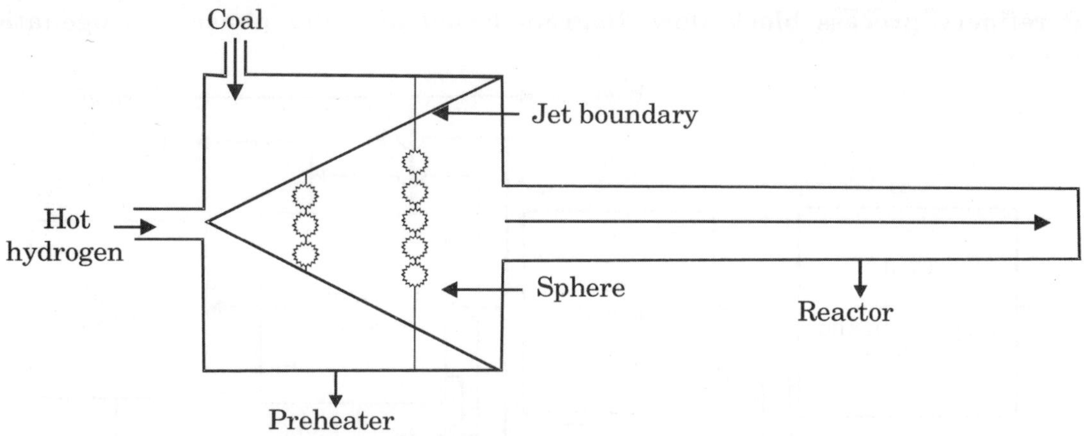

Figure 4.6

For the preheater type where one uses the technique of mixing due to turbulence by free jet, the velocity of jet varies in the radial direction; having lower velocity as one goes from jet axis towards the jet boundary. Thus fractions of coal which are away from the axis stay longer in the preheater while those near the axis stay shorter. Due to variation in heating time product distribution will vary as explained earlier. Unlike this preheater type, in the boundary layer separation technique there is no velocity gradient in the mixing zone behind the sphere and hence all the coal fractions stay same time in the preheater leading to one particular product distribution which depends on the contact time in the preheater only.

Apart from velocity variation in the turbulent mixing zone the degree of turbulence is also important because that determines how effective is the rapid heating. In the case of free jet, the turbulence due to entrainment is not so intense as in the case of boundary layer separation promoted turbulence behind the sphere.

Coal refinery process block flow diagram based on solid phase hydrogenation. (Fig. 4.7)

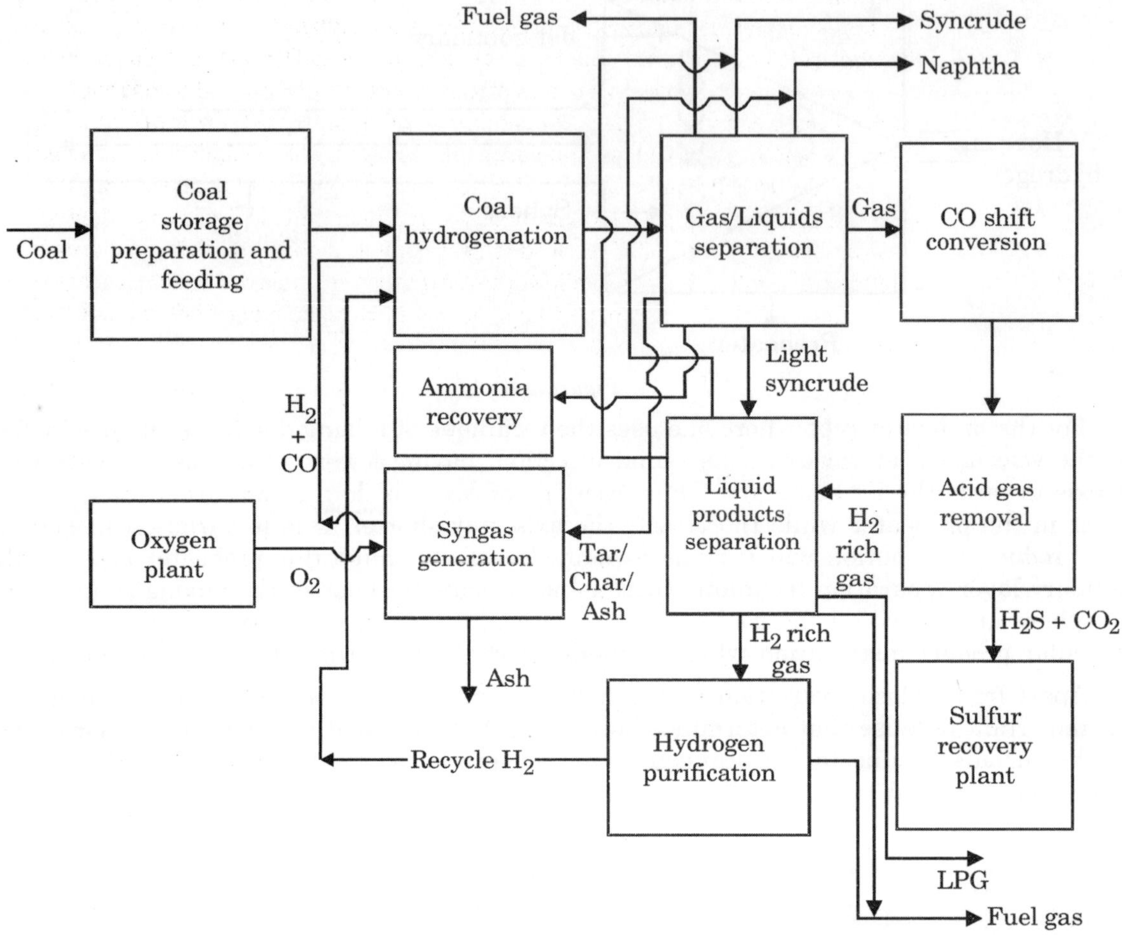

Figure 4.7

REFERENCES

1. ECN Petrochemicals' 79 Supplement, December 17, 1979.
2. W.C. Schroeder's US patent (1964) 3, 152, 063.
3. Indian Patent 208758 (2007) by Dr. Amalesh Sirkar.
4. Indian Patent 235876 (2009) by Dr. Amalesh Sirkar.
5. 'A kinetic model for solid phase hydrogenation of coal' by A. Sirkar, Indian Chemical Engineer, 45, No. 3, July–September, 2003.
6. 'Design of a novel preheater cum reactor for hydrogenation of coal' by A. Sirkar, S. Ghosh, S. Laha and M. Banerjee, Indian Chemical Engineer, 50, No. 1, Jan–March 2008, Corrsigendum – vol. 51, No. 2, April-June, 2009.
7. Personal communication with Dr. W.C. Schroeder.

Section I

CHAPTER 5

Hydrocracking of Heavy Bottoms to Gasoline and Other Products

Heavy oil cracking is a practical and profitable operation because of substantial yield of gasoline. In general there has been two major processes; one is catalytic cracking and the other one is non-catalytic cracking process. While catalytic cracking has been in operation for sometime the non-catalytic one has only been in operation in pilot plant scale and not commercialised yet. Here we shall discuss the non-catalytic process in details only as because a typical FCC unit's process description has already appeared elsewhere.

In the non-catalytic process[1] the heart of the plant is a special type of reactor where cracking takes place simultaneously with coke gasification.

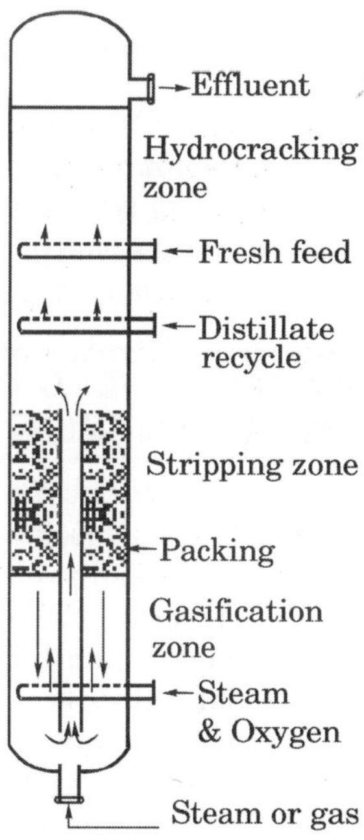

Figure 5.1

5.1 DESCRIPTION OF PROCESS AND REACTOR (REFER FIG. 5.1)

Feedstock is heated, fed to the reactor and cracked in presence of an inert carrier and a hydrogen containing gas in the upper bed. Part of the distillate product may be recycled to this hydrocracking zone. As hydrocracking occurs, coke is also formed and metals from the feedstock is deposited on the carrier.

The carrier continuously moves down through a stripping zone to the lower bed of the reactor where the coke is gasified with steam and oxygen to produce hydrogen rich gas. Distillate product may also be recycled to the gasification zone to increase the yield of fuel gas. The gas make flows upward through the annulus to the upper bed where the hydrogen provides the environment for the hydrocracking reactions.

The stripping zone serves two critical functions. By eliminating backmixing from the hydrocracking zone, the stripping section provides the needed temperature gradient between the hydrocracking and gasification zones. In addition, this zone allows the rising gas to strip absorbed liquid from the falling carrier so that coke alone is gasified in the lower bed.

The decoked carrier from the gasification section of the reactor is recirculated via a center transport tube using either steam or product fuel gas as the propellant. No unconverted resid feedstock remains and there is no product coke.

Hydrocracking zone temperature can be varied over the range of 925°–1400°F to produce a wide spectrum of product slates ranging from mostly liquids to nearly all fuel gas. The lower-bed gasification zone is operated between 1700°–1900°F to provide for gasification of the coke with steam and oxygen, the necessary heat for the hydrocracking zone and the desired water-gas shift.

Pressure of about 400 psig has been found to be desirable for maximizing liquid production. At this total pressure, sufficient hydrogen partial pressure is obtained to minimize diolefin formation. Total reactor pressure of about 600 psig is preferred when maximum fuel gas production is desired.

The above mentioned process was developed using a 6 in ID process development unit (PDU). The 55ft reactor was run for more than 10,000 hrs and processed as much as 15 bbl/day of feed. Continuous runs of upto 510 hr were made.

Typical yields of products from this reactor are as follows:

	Without recycle	With recycle
$C_1 - C_3$ (wt %)	12 – 17	11 – 19
C_4 – 400°F, vol %	28 – 45	34 – 56
400° – 700°F vol %	11 – 21	5 – 31
700°F$^+$ vol %	16 – 29	7 – 26
Carbon wt %	10 – 22	12 – 21

Normal operations showed no carrier agglomeration or loss of carrier movement or fluidity. No adherent deposits of coke or heavy liquids were found on the reactor internal. Readily available commercial material in an appropriate particle size distribution and porosity was found to be satisfactory for use as carrier.

5.2 DRAWBACKS OF THE ABOVE PROCESS AND SUGGESTIONS FOR IMPROVEMENT[(2)]

1. It is possible that carrier material may not pass uniformly through the packed bed due to blockage in some portions because of random packing arrangement. This may result in short circuiting of carrier material as well as the gas flowing upwards bypassing some portion of the bed. The disadvantage of this phenomenon is proper maintenance of uniform temperature gradient form bottom to top of packed bed in lateral direction thus resulting in variation in yield pattern. Such problem can be taken care of if we vibrate the reactor because the packing material will move laterally and this helps in free flow of carrier material downwards uniformly throughout the bed since the channels get opened up.

Vibrating the reactor will also help prevent interruption in smooth operation of the reactor due to portion of carrier material filling up portion of the bed and not being able to flow downward due to channel blockage. This can happen due to random filling of packing material forming a non-uniform packed bed.

2. The other drawback of the reactor design is the method of feeding the hot bottoms into the reactor (in the cracking zone). Figure 5.1 shows perforated tubes for feeding the hot bottoms. The disadvantage of this design is that only carrier material close to the tube will receive the oil on it with a thick layer of it and the carrier material in the lower portion will run dry. A thick layer of oil means more diffusional resistance for hydrogen to reach the oil layer close to carrier material leading to excessive coke formation. The other disadvantage of the existing feeding technique is that the perforated tubes within the bed change the fluidization characteristics of the bed because the tubes act as baffles within the bed. This drawback will be more so especially in the commercial size reactor which is much bigger than the pilot plant where there will have to be number of tubes for equal distribution of hot bottom.

   A proper design would be to in stall spray nozzles at different levels of the cracking zone (on the wall of the reactor) so that all the carrier material receives oil uniformly. Also, this is facilitated by a smaller reactor diameter. A larger diameter means the central zone running dry due to interference by carrier material close to the wall. This can be avoided if we also install nozzles on a feed pipe located centrally within the cracking zone.

REFERENCES

1. The Oil and Gas Journal, August 8, 1977.
2. Indian Patent 200688 (2007) by Dr. Amalesh Sirkar.

Section I

CHAPTER 6

Naphtha Cracking

The traditional method[1] of production of ethylene and coproduct is by the pyrolysis of naphtha. Feedstock is preheated through various process heat exchangers and then passed through the convection section of a furnace whereby the temperature goes up to 300–400°C. Here the feedstock is mixed with steam (700–800°C) and routed into the radiant section of a turbular furnace. Eventually the feedstock is heated to 800–850°C, the furnace temperature being higher than 900°C. However, if one can heat naphtha quickly (the time being in the order of hundredth of a second) to the desired temperature then one can have a better control over the product distribution leading to higher yield of propylene and ethylene. This is because in the conventional process cracking begins at a lower temperature while heating is continued. Quick heating to higher temperature means higher value of rate constant to begin with leading to higher yield and less volume of the cracker due to higher initial rate of reaction.

6.1 PROCESS DESCRIPTION

If one can do direct heating of naphtha using hot solid particles through rapid mixing technique then such a process might be feasible. The inert solids can be circulated between the preheater-cum-cracker of special design having short residence time preheater and solids preheater. In such designs we do not require complicated coil design as required in previous designs and effective mixing between solid and gas can cut down the volume of the preheater substantially. For the moving bed type design, solids have to be separated from gas. For this, after the short residence time preheater one can introduce the solid and gas mixture into a separator-cum-cracker where solids will come out from bottom and gases will come out from side outlet. The separator has to give the desired reaction time for conversion of naphtha to desired products. After this gravity separator one can introduce quenching oil to stop the reaction at the exit of the separator. Figure 6.1 shows the process scheme for the case where preheater is like a moving bed.

Alternatively, if one uses a preheater design such as a vibrated bed where solids move up and down, then one will have to handle two streams; either the preheated naphtha vapour coming out from top or from side and solids from bottom. In such a case quickly heated naphtha vapour from top or side can pass through a cracker (well insulated) where only endothermic heat of cracking will be supplied by the naphtha vapour itself and in the process vapour will get cooled somewhat (adiabatic cracker). Alternatively, the cracker portion can be made like a furnace to supply the heat of cracking thus maintaining higher reaction temperature which would require less volume of cracker for desired yield.

Solids from the bottom of vibrated bed has to flow back to the solids preheater where fuel oil or natural gas could be used to preheat the solids. Figure 6.2 shows the process scheme.

Unlike conventional cracking process where steam is used to react with carbon deposited on the wall of the cracker, in the present process, steam may not be used in the cracker since residence time in the cracker portion is much less than conventional cracker leading to minimum amount of carbon formation. With minimum carbon formation the period of operation of the cracker can be extended in comparison to conventional cracker before cleaning cycle.

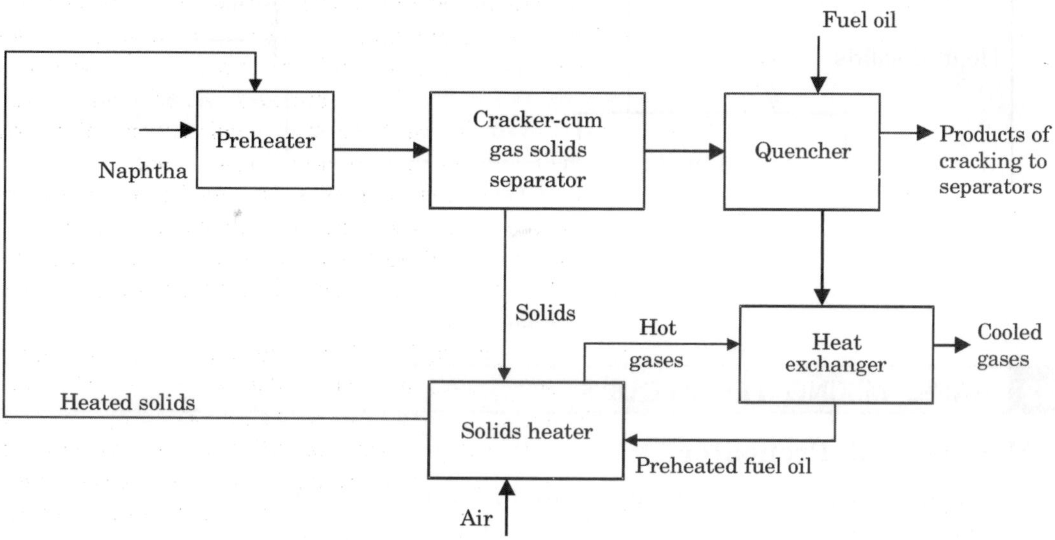

Figure 6.1

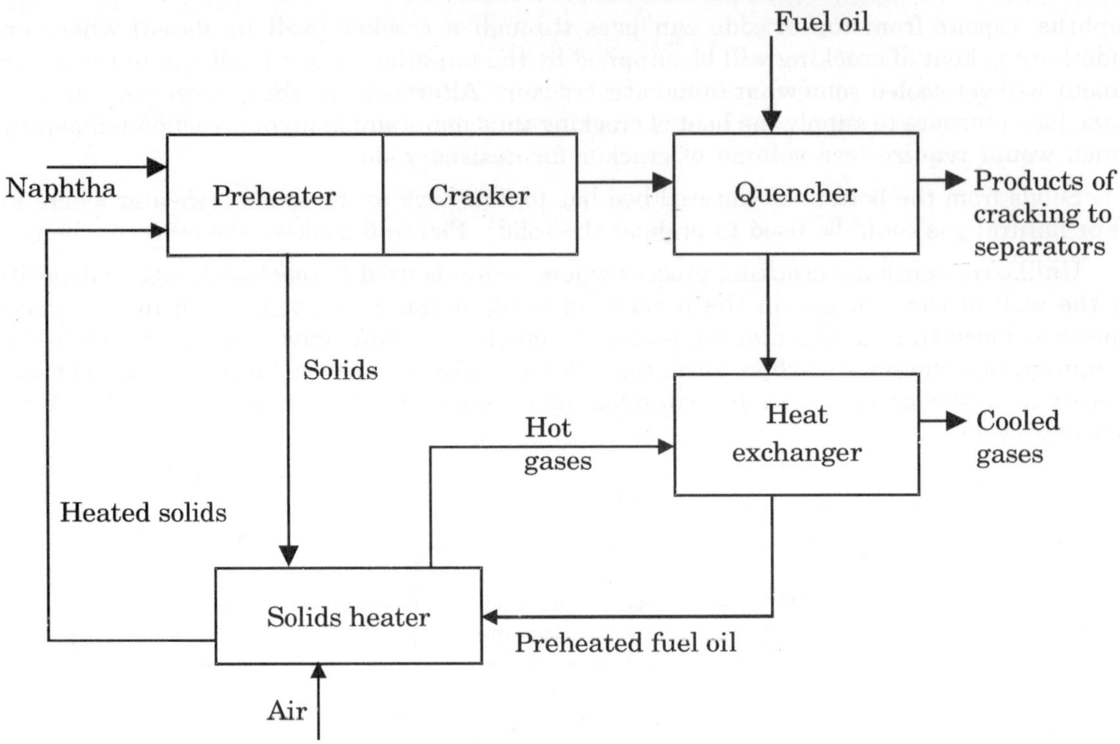

Figure 6.2

6.2 RAPID MIXING TECHNIQUES

1. Moving Bed Preheater

Use short residence time preheater used for coal liquefaction (chapter 2 section II). The only difference being that in the present case solids heat the gas whereas in the case of coal liquefaction gas (H_2) heats solids (coal particles). If we are using the same short residence time preheater as used for heating coal quickly to operating temperature, then it would do the reverse heating efficiently as in the present case of naphtha preheating provided either we use particles of similar size, specific gravity, similar specific heat and similar thermal conductivity as those of the bituminous coal used for coal liquefaction or we change the parameters. The reasons are as follows:

(*a*) If the specific gravity of the particles used is higher than coal then for mixing of gas and solid, linear velocity of naphtha vapour has to be high such that particles will not move downward due to its weight and thereby avoid mixing.

(*b*) If thermal conductivity is not similar then heat transfer rate will be dissimilar requiring more time for heating naphtha vapour if thermal conductivity is lower than coal due to change in overall heat transfer coefficient. Hence, in such a case cracking reactions will start within the short residence time preheater due to longer preheating time resulting in different yield pattern giving less ethylene as in the conventional naphtha cracker. If thermal conductivity is higher then naphtha would be heated quickly.

(c) If specific heat of solids is less than that of coal then naphtha may be heated quickly since circulation rate of particles is large having larger surface area overshooting the lower heat transfer coefficient. On the other hand, if specific heat is larger, then circulation rate will be smaller giving less surface area of particles thus taking longer to preheat naphtha provided heat transfer coefficient is not large. If the solids' particle size is small like coal particles (200 mesh) a gravity separator may be appropriate since the specific gravity of solids would be high (*e.g.*, steel) such that settling velocity would be appropriate for quick settling thus requiring small residence time in the separator for cracking reactions to occur effectively.

Design procedure for estimating volume of moving bed preheater is elaborated in chapter 2 (section II) under the heading 'Reactor Design'.

2. Vibrated Bed

(a) Solids Moving up and Down and Naphtha Moving from Bottom to Top

If one uses solid particles of the shape shown in Fig. 6.3, preheat these in a moving bed solids preheater by burning fuel oil or natural gas and allow the particles to pass through a vibrated preheater (Fig. 6.5) then naphtha can be preheated quickly by upward and downward movement of the particles. For instance, during upward and downward movement of the particles naphtha vapour will be displaced sidewise in the form of waves due to compressive action of the particles. Besides that, due to curvature of the particle towards the joint the waves would move towards the joint and collide there undergoing mixing within the wave fronts. Collision between the waves occurs due to flimsy surface of the wave. After collision mixing within the gaseous medium occurs due to the following phenomena:

Note that the waves coming from opposite direction are in a curvilinear path and results in shearing action due to low viscosity of gases as they penetrate each other, hit the joint and bounce upward as shown below (Fig. 6.4). Due to the movement of the gaseous layers from bottom to top there occurs a circulatory motion within the gaseous medium which destroys the temperature gradient that was present in the wave prior to collision.

Because wave front has a temperature gradient from surface temperature of particle to bulk temperature of gases due to gas-solid contact, the mixed temperature of the gases is the average of the two assuming linear variation in temperature across the thickness of wave. The mixed naphtha vapour will be continuously swept away in the vertical direction by the upward moving naphtha vapour, thus raising its temperature. The solids after loosing heat will come out from the bottom as shown in Fig. 6.5. The upward movement of the solids is achieved by means of crank shaft arrangement attached to a guide block which hits the preheater intermittently.

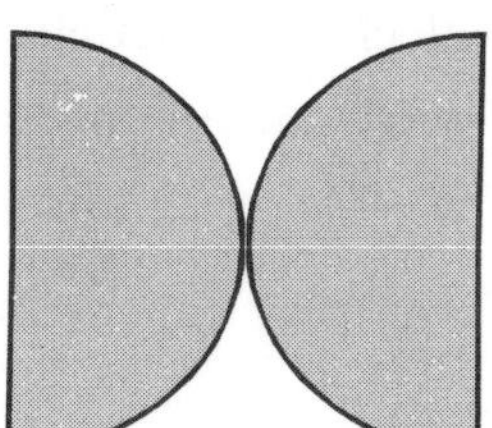

Figure 6.3: Particle shape

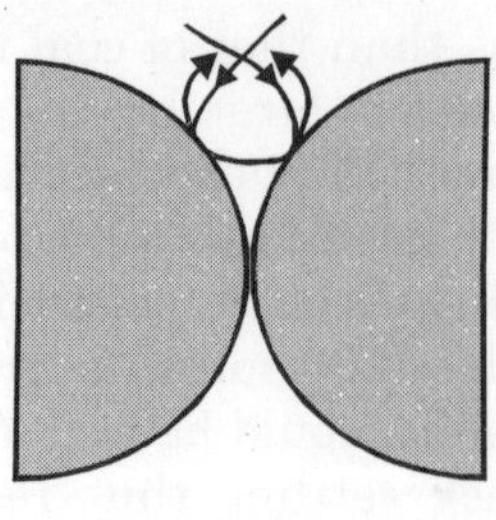

Figure 6.4: Mechanism of mixing

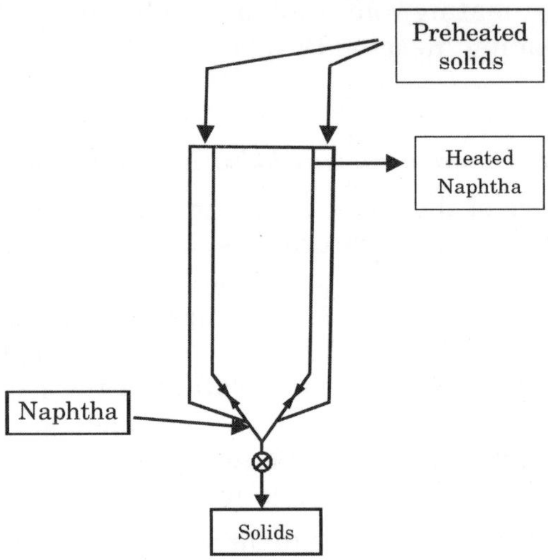

Figure 6.5

Mathematically, the heating phenomena can be expressed as follows:

1. Momentum Balance Equations

$$N_c \times A_c \times dy \times W_p \times V_{p_y} = N_c \times A_c \times dy \times A_p \times y_f \times V_{wy} \times \rho_{yv} \quad ...(1)$$

$$N_c \times A_c \times dy \times W_p \times V_t = N_c \times A_c \times dy \times A_p \times y_f \times V'_{wy} \times \quad ...(2)$$

where V_{p_y} – upward velocity of particles at y

V_t – terminal velocity of particles

V_{wy} – velocity of wave moving towards the joint during rise of the particle

V'_{wy} – velocity of wave moving towards the joint during fall of the particle

ρ_{yv} – density of naphtha vapour at y

2. Heat Balance Equation

$$W_{Nap} C_{p_{Nap}} dt = N_c \times A_c \times dy \times A_p \times (V_{wy} + V'_{wy}) \rho_{yv} \times C_{p_{Nap}} [(t + 925)/2 - t] \quad ...(3)$$

Note that solids enter at 950°C and leave at 900°C so that the average solid temperature is 925°C.

or $$W_{Nap} C_{p_{Nap}} \, dt = \frac{N_c \times A_c \times dy \times W_p \times (V_{py} + V_t)}{y_f} \times C_{p_{Nap}} \times [(t + 925)/2 - t] \quad \text{...(4)}$$

$$V_{py} = V_{p_o} \; \tilde{n} \; g \; \theta_{up} \text{ where } \theta_{up} = \frac{y}{V_{py}}$$

$$= V_{p_o} - \frac{gy}{V_{py}}$$

$$V_{py}^2 - V_{p_o} V_{py} + gy = 0$$

$$V_{py} = f\left(V_{p_o}, y\right) \quad \text{...(5)}$$

$$\therefore \quad \int_{t_0}^{850°C} \frac{dt}{(t+925)/2 - t} = \frac{N_c \times A_c \times W_p}{W_{Nap} \times y_f} \int_0^L (V_{py} + V_t)\, dy \quad \text{...(6)}$$

$$W_{solids} \times C_{p\ solids} \times 50 = W_{Nap} \times C_{p_{Nap}} \times (850 - t_0) \quad \text{...(7)}$$

Assuming heating time to be 0.01 sec

(a) Solids holdup

$$W_{solids} \times 0.01 = N_c \times A_c \times L \times W_p \quad \text{...(8)}$$

Also, $$A_c \times dy = \frac{W_{Nap}}{\rho_{yv}} d\theta \quad \text{...(9)}$$

$$\int_0^{\theta_T} d\theta = 0.01 = \frac{A_c}{W_{Nap}} \int_0^L \rho_{yv}\, dy$$

or $$A_c = \frac{W_{Nap} \times 0.01}{\int_0^L \rho_{yv}\, dy} \quad \text{...(10)}$$

equation (6) becomes (after substitution from (8))

$$\int_{t_0}^{850°C} \frac{dt}{(t+925)/2 - t} = \frac{W_{solids} \times 0.01}{W_{Nap} \times y_f \times L} \int_0^L (V_{py} + V_t)\, dy \quad \text{...(11)}$$

The height of the preheater 'L' can be estimated by the following equation

$$2\int_0^L \frac{dy}{V_{py}} + \frac{2L}{V_t} = 0.01 \text{ sec} \quad \text{...(12)}$$

Since the solid particles are continuously fed to the preheater at bottom, during each cycle of upward and downward movement the inventory of solids builds up which move up and down during the next cycle. Hence the factor '2' in the above equation. The solids' entry and discharge occur through valves operated by timer.

For a given V_{p_0} and V_t having 'L' estimated from equation (12), it can be used to estimate 'y_f', the wave front thickness from equation (11). Note that 'y_f' varies with variation in V_{p_0} and V_t which is logical because the compressive action of the particles on naphtha vapour for formation of wave varies due to variation in particle velocity. 'y_f' also depends on weight per particle W_p, because thrust by the particle is proportional to (velocity × weight).

W_{solids} is estimated from equation (7).

A_c is obtained from equation (10) since $\rho_{vy} = f(t)$ and taking a mean value of ρ_{yv}.

N_c is estimated from equation (8) knowing A_c, L and $W_{solids.}$

Note that W_p (weight per particle) depends on particle shape, size and material of construction. Particle shape is shown in Figure 6.3.

Left hand side equation (11) may be lower or higher depending on the final temperature after preheating. So experiments should be carried out using different particle shape and size and different V_{p_0} to determine 'y_f'. Next, y_f should be correlated to V_{p_0}, V_t and W_p after which proper value of these variables are to be choosen such that the y_f calculated based on the correlation matches L.H.S. of equation (11) corresponding to final preheating temperature of 850°C. The suggested correlation for y_f may be as follows:

$y_f = b\,(W_p \times V_t)^m \times (W_p \times V_{p_0})^n$ because higher the thrust, higher is y_f due to greater compressive action.

Note that for L.H.S. of equation (11) to be larger such that preheating temperature is higher, in the R.H.S. $\dfrac{\int_0^L (V_{p_y} + V_t)\,dy}{y_f \times L}$ should be large. Since y_f is directly proportional to D_p, V_{p_0} and V_t, because it depends on thrust by the particle and since V_{p_0} and V_t also appear in the numerator, for the ratio to be large D_p, V_{p_0} and V_t should be choosen properly, such that the ratio corresponds to L.H.S. value with final preheating temperature of 850°C. It may be that a combination of optimum particle size and V_{p_0} exists which would result in the desired final temperature.

Since the particles are not spheres, while lifted, these may wabble and hence the contact at the top wall may be at an angle such that while dropping down the particles may hit the side wall and bounce back. In such a case wave formation may not be proper, leading to less effective preheating. Thus, it may be desirable to enlarge the cross-section downwards *i.e.,* tapered such that particles may not hit the side wall while dropping down.

The above mentioned phenomenon will be the situation when particles hit the top wall at a positive velocity. But, if the particles hit the top wall at zero velocity then they will fall vertically and will quickly attain terminal velocity. In this case the height of the preheater should be such that it attains zero velocity at the top wall with proper lifting velocity V_{p_o}. Both the designs could be tried and checked out as to which one is better in terms of quick heating of naphtha vapour.

(b) Solids Moving Up and Down and Naphtha Moving Crosswise[(4)] (Fig. 6.6)

For better heat pick up of mixed wave by the naphtha vapour from the surface of the particles at any level one can install a multiple concentric pipe such that through the inner pipe naphtha enters the preheater and is dispersed laterally through several openings (holes) at any height of the preheater. The solids move up and down due to vibratory movement as explained in case 2 (*a*). The dispersed vapour after picking up mixed waves from all the particle surfaces will travel outward from the preheater outer wall through the openings and travel through a conical cracker as shown in Fig. 6.6. The reason for conical shape cracker is to achieve high velocity such that the flow regime is turbulent (approaching plug flow) to maximize conversion to desired products. For case (*a*) cracker is a separate unit connected to preheater unlike case 2 (*b*) where preheater and cracker is a combined unit. Solids enter the annular space at bottom after each cycle of upward and downward movement of particles and undergo another cycle and so on. The solids' entry and discharge occur through valves operated by timer.

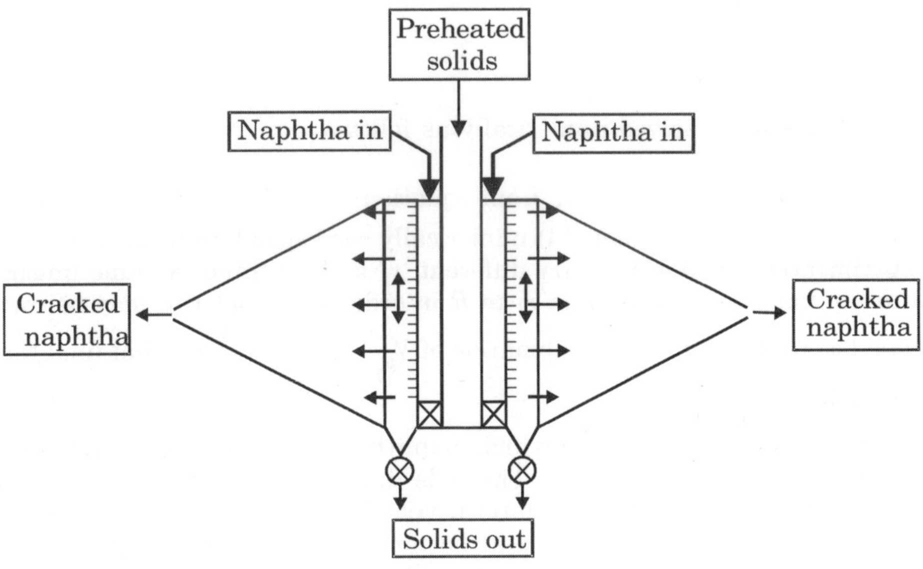

Figure 6.6

Mathematically, the heating phenomena can be expressed as follows:

1. In this case same momentum equations (1) and (2) are valid.
2. Heat Balance Equations:

$$W_{Nap} C_{p_{Nap}}\, dt = \frac{N_c \times 2\pi r dr \times dy \times W_p}{y_f} \times (V_{py} + V_t) \times C_{p_{Nap}} \times \left[\left(\frac{t+925}{2}\right) - t\right] \qquad ...(13)$$

Integrating,

$$\frac{\Delta t}{\left[\left(\frac{t+925}{2}\right)-t\right]_{av}} = \frac{N_c \times 2\pi \times \left(\frac{r_1+r_2}{2}\right)\Delta r \times W_p}{W_{Nap} \times y_f} \int_0^L (V_{py}+V_t)\,dy \qquad \text{...(14)}$$

$$\Delta r = r_2 - r_1$$

From equation (7) one gets W_{solids}. For V_{p_y} use equation (5)

<u>Solids holdup</u>

$$W_{solids} \times 0.01 = N_c \times (\pi R^2 - \pi r_1^2) \times L \times W_p \qquad \text{...(15)}$$

Substitute in equation (14) for N_c from this equation and then equation (14) is dependent on 'L' and 'R' only.

Also,

$$2\pi r dr \times L = \frac{W_{Nap}}{\rho_{rv}} d\theta$$

$$\theta_T = \int_0^{\theta_T} d\theta = \frac{L}{W_{Nap}} \int_{r_1}^{R} 2\pi r\, \rho_{rv}\, dr = 0.01 \qquad \text{...(16)}$$

$\rho_{rv} = f(t)$ and $t = \phi(r)$

Equation (14) is to be solved numerically as follows:

For a given V_{p_0} and V_t estimate 'L' from equation (12). *Calculate R from equation (16).* Next, assume 'y_f' and solve equation (14) numerically corresponding to the final temperature obtained experimentally. If not, then try different 'y_f' and continue. Assume linear variation of temperature of naphtha vapour from r_1 to R in order to get $\Phi(r)$. y_f may be correlated as before and one has to choose proper combination of V_{p_0}, V_t and W_p such that final temperature of 850°C is obtained.

r_1 depends on the velocity with which naphtha enters the preheater which is the header pipe and then travels laterally. The hole diameters of the concentric pipe should vary from top to bottom such that the lateral velocities through the holes are different– smaller diameter at top and larger holes at bottom so that the flow is uniform along the cross-section at all heights of the preheater. Pressure is higher at top and lower at bottom. Due to higher pressure at top and smaller holes, velocity is higher at top and lower at bottom.

3. (a) Countercurrent Movement of Naphtha Vapour and Solid Particles

Here, naphtha vapour enters the vertical preheater at bottom and moves upward while the particles undergo free fall from top of the preheater. Mechanism of heating naphtha vapour is same as described under Section 2(*a*). Mathematically, the heating phenomenon can be expressed as follows:

1. Momentum Balance Equation:

$$N_c \times A_c \times dy \times W_p \times V_t = N_c \times A_c \times dy \times A_p \times y_f \times V'_{wy} \times \rho_{yv} \quad ...(17)$$

2. Heat Balance Equation:

$$W_{Nap}\ C_{p_{Nap}}\ dt = N_c \times A_c \times dy \times A_p \times V'_{wy} \times \rho_{yv} \times C_{p_{Nap}} [(t + t_s)/2 - t] \quad ...(18)$$

or $$W_{Nap}\ C_{p_{Nap}}\ dt = \frac{N_c \times A_c \times dy \times W_p \times V_t}{y_f} \times C_{p_{Nap}} \times [(t + t_s)/2 - t] \quad ...(19)$$

$$W_{solids} \times C_{p_{solids}} \times 50 = W_{Nap} \times C_{p_{Nap}} \times (850 - t_0) \quad ...(20)$$

∵ solids exit at 900°C and enter at 950°C.

Also,

$$W_{solids} \times C_{p_{solids}} \times (950 - t_s) = W_{Nap} \times C_{p_{Nap}} \times (t - t_0) \quad ...(21)$$

From (20) $W_{solids} = \dfrac{W_{Nap} \times C_{p_{Nap}} \times (850 - t_0)}{50 \times C_{p_{solids}}}$

And from (21) $950 - t_s = \dfrac{W_{Nap} \times C_{p_{Nap}} \times (t - t_0)}{W_{solids} \times C_{p_{solids}}}$

or $$t_s = 950 - \frac{\not{W}_{Nap} \times \not{C}_{p_{Nap}} \times (t - t_0) \times 50}{\not{W}_{Nap} \times \not{C}_{p_{Nap}} \times (850 - t_0)}$$

or $$t_s = 950 - \frac{50(t - t_0)}{850 - t_0} \quad ...(22)$$

Substitute the above expression for t_s in (19)

$$\therefore\ W_{Nap}\ dt = \frac{N_c \times A_c \times W_p \times V_t}{y_f} \times \left[\left\{\left(t + 950 - \frac{50(t - t_0)}{850 - t_0}\right)/2\right\} - t\right] dy$$

or $$\int_{t_0}^{850} \frac{dt}{\left[\left\{\left(t + 950 - \frac{50(t - t_0)}{850 - t_0}\right)/2\right\} - t\right]} = \frac{N_c \times A_c \times W_p \times V_t}{W_{Nap} \times y_f} \int_0^L dy \quad ...(23)$$

From equation (8) $N_c \times A_c \times W_p = \dfrac{0.01 \times W_{solids}}{L}$

∴ R.H.S. of equation (23) becomes

$$\frac{W_{solids} \times 0.01 \times V_t}{W_{Nap} \times y_f}$$

The height of the preheater 'L' is given by

$$\frac{L}{V_t} = 0.01 \quad \text{...(24)}$$

For a given particle shape and size V_t is known and hence 'L' can be estimated from the above equation.

W_{solids} is estimated from equation (20)

A_c is estimated from equation (10)

Choosing a particle shape, size and material V_t becomes known and from experiment one determines the final temperature and hence, y_f becomes known from equation (23). This way y_f may be correlated to V_t and W_p and using this correlation one can determine what V_t and W_p correspond to 850°C temperature from equation (23).

Comparing equation (12) and (24) one can conclude that for a given particle shape and size, height of preheater is less with vibrated bed in comparison to non-vibrated bed, since terminal velocity 'V_t' is same in both the cases. However, vibrated bed preheater is costlier than non vibrated bed because of the additional design features as follows:

1. Vibratory Mechanism with power consumption
2. Larger dia preheater because of solids entry section surrounding the main heater
3. Requirement of timer operated valves

3. (b) *Solids Moving Down and Naphtha Moving Crosswise*

Mathematical Model of the Heating Phenomenon

1. Equation (17) is valid for momentum balance
2. Heat Balance Equations:

$$W_{Nap} C_{p_{Nap}}\, dt = \frac{N_c \times 2\pi r dr \times dy \times W_p}{y_f} \times V_t \times C_{p_{Nap}} \times [(t + t_s)/2 - t] \quad \text{...(25)}$$

Equation (20) is valid here.

Since the soild temperature is varying from 950˚C to 900˚C from top to bottom one can take t_s to be the average temperature 925˚C for simplicity purpose.

Solids Holdup

Equation (15) is valid.

Equation (25) on integration becomes

$$\frac{\Delta t}{\left[\left(\frac{t+925}{2}\right) - t\right]_{av}} = \frac{N_c \times 2\pi \times \left(\frac{r_1 + r_2}{2}\right) \Delta r \times W_p}{W_{Nap} \times y_f} \times V_t \times L \quad \text{...(26)}$$

Equation (16) is also valid.

Equation (26) is to be solved numerically as follows after substituting for N_c from equation (15):

For a given particle size and shape V_t is known and 'L' is obtained from equation (24). *Calculate R from equation (16).*

Next, assume 'y_f' and solve (26) numerically corresponding to the final temperature obtained experimentally. If not, then try different 'y_f' and continue as in case 2(*b*). Also, choose proper V_t and W_p using the correlation for y_f as before such that desired temperature of 850°C is obtained.

6.3 ALTERNATIVE PARTICLE SHAPE (FIGURES 6.7, 6.9 AND 6.10)

1. It is like a star with several arms and the arms have curved cross-section. For the star shaped particles having arms with curved cross-section, waves are generated by adjacent arms due to compressive action of the particles on the naphtha vapour during upward and downward movement of the particles. The adjacent waves follow a curvilinear path due to curved cross-section of the arms. The waves collide with each other due to flimsy surface of the waves. After collision they penetrate each other due to low viscosity of the vapors and move inward taking shortest path due to resistance offered by the gaseous medium and encounter each other in the middle region. After the gaseous layers encounter each other circulatory motion gets generated in upward and downward direction due to least resistance offered in vertical direction for further movement. Note that in horizontal direction mixing does not help because they are of same temperature from both direction. Only circulatory motion in vertical direction helps proper mixing between lower and higher temperature layers of the vapour in cocurrent/countercurrent movement, countercurrent or cross current movement of vapour as the case may be. In the case where vapour moves from bottom to top the flow has to be turbulent such that enough mixing occurs between the rising vapour and the collided waves, thus raising bulk temperature of the rising vapour. For cross current flow of vapour an optimum velocity of vapour is needed such that wave formation occurs without much of interference by the mainstream vapour since both moves laterally and at the same time just enough turbulence occurs by the mainstream to mix with collided waves.

   In the previous particle shape suggested (Fig. 6.3) heat pick up is not that efficient because wave collision occurs at the joint of each particle (on its surface) and naphtha moves through the void space and comes in contact with mixed wave of naphtha vapour only at the periphery of the particle. For cross flow of naphtha vapour this particle is also better because in the previous shape heat pick up is better only if the particles are in a particular orientation as shown below (Fig. 6.8). For star shaped particle it exists only in a horizontal plane as it rises.

Figure 6.7: Star shaped particle

2. The disadvantage of star shaped particle is that at a larger distance from the centre the adjacent waves may not get a chance to collide before the rising vapour comes in contact, since the waves have to travel longer distance laterally in order to collide with each other.

   Similarly, in cross current flow, vapour velocity has to be low enough for giving a chance to the adjacent waves to collide at a larger distance from the centre of the particle, which may not be turbulent enough for proper mixing to take place. To avoid this one may use a particle in the shape of a square made up of thin curved strips with space between the strips (Fig. 6.9). In such a case the gap between the strips is uniform and hence one can avoid the disadvantages mentioned above with star shaped particle. However, square shaped particle may not be as good as star shaped particle from other point of view *i.e.*, such particle while rising may wabble more due to momentum imbalance near the edges thus preventing uniform wave formation. Thus, modification of star shaped particle in which the arms are tapered outward (Fig. 6.10) such that there exists uniform clearance between the arms is the answer to an improved shape. The waves generated by the tapered arms will undergo collision and mixing according to the same principle described above for star shaped particle having curved arms. For this particle shape the wave has a variable area (but uniform thickness) because of the geometry of the arm, which is tapered. The gap between the arms should be such that the adjacent waves undergo collision in totality (from centre of the particle to its periphery) above or below the gap during downward and upward movement of the particles respectively. Actually, the gap dimension should be equal to the half of the width of the arms (at the outer edge) to make sure that for the wave portions at the end of the arm (periphery) undergo full collision in the space between two arms such that the rising vapour creates turbulence in the gap and picks up the mixed wave and thus its bulk temperature rises.

6.4 OPTIMUM PARTICLE SIZE (REFER EQUATION 8)

For a given naphtha flow rate $A_c \times L$ is nearly same whether we choose a small or large particle since

$$W_{Nap} \times 0.01 = A_c \times L \times \rho^{Nap}_{mean}$$

∴ If particle size is large W_p will be large and N_c will be less where N_c is no. of particles per unit volume of preheater and W_p is weight per particle. Two things are to be considered in this respect as follows:

1. Bed voidage will be different if particle size is different
2. If N_c is different then also bed voidage changes

Due to (1) and (2) linear velocity of naphtha will be different and depending on the particle Reynold's No. : the degree of turbulence created by the vapour to mix with the collided waves will be different. Besides this, N_c has to be optimum because there should not occur interference between the particles which may prevent uniform wave formation.

Particle Shapes to be choosen: ***<u>Star shaped</u>***

1. rectangular arms
2. tapered arms
3. square shape with rectangular strips

High naphtha velocity means more pressure drop but greater turbulence and low velocity means less pressure drop but less turbulence. So one has to choose optimum particle size for minimum pressure drop but complete mixing of rising vapour with collided waves.

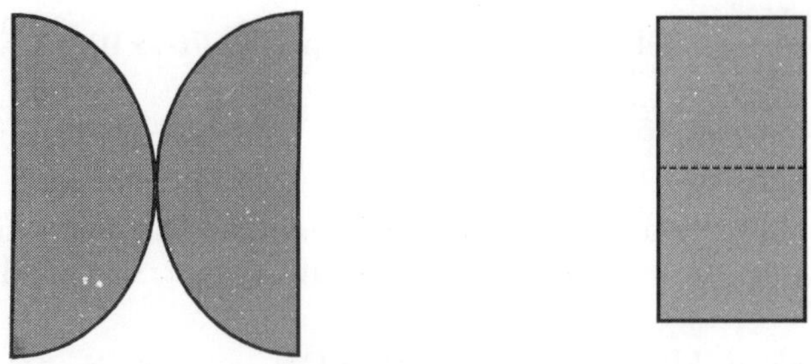

Figure 6.8: Preferred orientation Unfavourable orientation

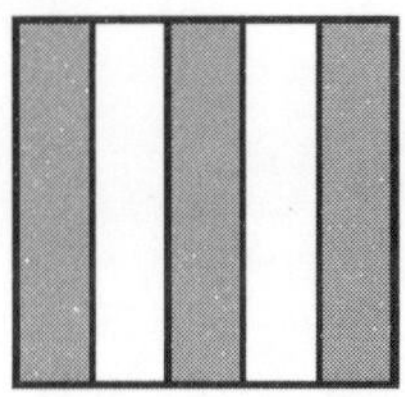

Figure 6.9: Square shaped particle

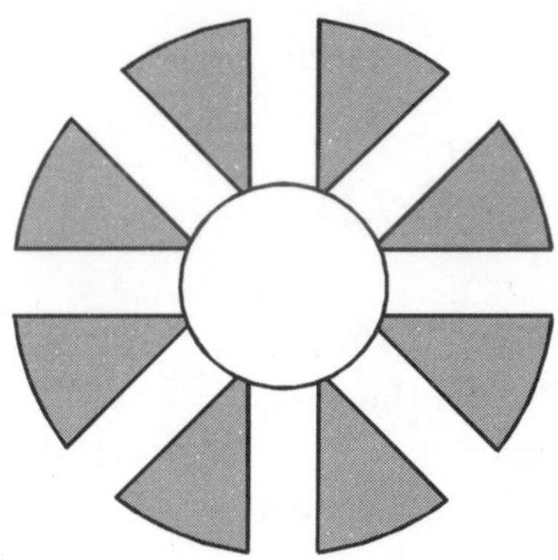

Figure 6.10: Modified star shaped particle

6.5 COMPARISON OF PREHEATER DESIGNS WITH FLUIDIZED BED PREHEATER

Volume of Preheater

Case I. Refer the design 2(*a*) in the text

Basis : 10 Kg/hr feed gas

From equation (12) for zero velocity at the top wall of preheater, $\frac{V_{p_o}}{g} + \frac{L}{V_t} = 0.005$

Assume V_{p_o} = 2 cm/sec

Consider particle shape shown in Fig. 6.3

Radius of circle = 0.15 cm

Width of particle = 0.15 cm

$$\therefore \quad D_p = 6 \times \frac{(0.15)^2}{0.3} = 0.45 \text{ cm}, \rho_p = 7.8 \frac{\text{gm}}{\text{cc}}, \rho_{av} = 1.767\times10^{-3} \frac{\text{gm}}{\text{cc}}$$

$$\therefore \quad V_t^{(2)} = 1.75\sqrt{\frac{981\times0.45\times7.8}{1.767\times10^{-3}}}$$

Instead of a typical naphtha for which we do not have all the properties, we have assumed a typical hydrocarbon say *n*-octane to verify the preheat concept, whose MW and BP fall within the range of naphtha cut.

V_t = 2443 cm/sec, MW = 114, BP = 126°C

Re_p = 14016 which is > 1000 and < 200,000

$$\therefore \quad \frac{L}{V_t} = 0.005 - \frac{2}{981} = 2.96\times10^{-3}$$

or L = 7.23 cm.

From equation (10)

$$A_c = \frac{W_{Nap}\times0.01}{\int_0^L \rho_{vy}\,dy}$$

$$= \frac{10\times1000}{3600}\times\frac{0.01}{1.767\times10^{-3}\times7.23}$$

$$= 2.17 \text{ cm}^2$$

∴ Volume of preheating chamber = 15.7 cc

Case II. Refer the design 3(*a*) in the text

$$\frac{L}{V_t} = 0.01 \qquad \therefore \quad L = 24.43 \text{ cm.}$$

A_c = 0.64 cm^2

∴ Volume of preheating chamber = 15.7 cc

Case III. Fluidized Bed Preheater

$$\frac{V_t^{(2)}}{\overline{V}_{OM}} = \frac{2.32}{\epsilon_M^{3/2}}$$

Assume ϵ_M = 0.5

$$\therefore \quad \overline{V}_{OM} = \frac{2443\times0.3535}{2.32}$$

$$= 372.3 \text{ cm/sec}$$

R_{e_p} = 2136 > 1000 but less than 200,000

We shall use the following correlation for calculation of heat transfer coefficient[3]

$$h = 1.77 \times c_p \times G \times \left(\frac{C_p\mu}{\Re}\right)^{-2/3} \left(\frac{D_p G}{\mu(1-\in_M)}\right)^{-0.44}$$

$$G = V \times \rho_{av}$$

$$= \frac{372.3 \times 1.767 \times 10^{-3} \times 3600}{453.4} \times (30.48)^2$$

$$= 4853 \frac{lbs}{(hr)(ft^2)} = \frac{0.658\text{gm}}{\text{cm}^2\text{ sec}}$$

$$\frac{D_p G}{\mu(1-\in_M)} = \frac{0.45 \times 0.658 \times 100}{0.01386 \times .001 \times 1000 \times 0.5}$$

$$= 4272.7 \text{ which is} > 30$$

$$\frac{C_p\mu}{\Re} = \frac{0.783 \times 0.01386 \times .001 \times 1000}{100 \times 1.83 \times 10^{-3}} \times 418.7$$

$$= 24.8$$

$$C_{P_{Mean}} = 0.783 \frac{Cal}{gm°C} \qquad \mu_{Mean} = 0.01386 \text{ cp}$$

$$\Re_{Mean} = 1.83 \times 10^{-3} \frac{W}{m°C}$$

$$\therefore \quad h = 1.77 \times 0.783 \times 4853 \times (24.8)^{-2/3} (4272.7)^{-0.44}$$

$$= \frac{1.77 \times .783 \times 4853}{39.58 \times 8.5} = 20 \frac{Btu}{(hr)(ft^2)(°F)}$$

$$\text{LTMD} = \frac{(900-126)-(950-850)}{\ln \frac{774}{100}} \times \frac{9}{5}$$

$$= 593°\text{F}$$

∴ Surface area of particles required

$$= \frac{Q}{h\,(LMTD)} = \frac{10 \times 1000}{453.6} \times \frac{0.783 \times (850-126)}{20 \times 593} \times \frac{9}{5} = 1.9 \text{ sq.ft}$$

$$\text{No. of particles in the bed} = \frac{1.9 \times (30.48)^2}{3.14 \times (0.30) \times 0.15} = 12492$$

Volume of each particle = 0.0106 cc

$$\therefore \text{ Bed volume} = \frac{12492 \times .0106}{1-\in_M} = 265 \text{ cc}$$

∴ Compared to case I and case II Fluidized Bed Preheater Volume required is much higher.

6.6 GAS–SOLID SEPARATION FOR MOVING BED PREHEATER

Gravity Separator-cum-Cracker (Fig. 6.11)

(*a*) Vapour Velocity Profile in the Separator.

(*b*) Ethylene Production Rate and dimensions of Separator.

(*a*) Since the solids retard the movement of vapour the velocity profile in the separator is as follows : In the z direction vapour flow is obstructed by the solid particles which create drag on the vapour and hence the velocity is reduced at $y = -H/2$ where solids density is highest in the separator. At the top wall of separator velocity is zero. Maximum velocity is near the top wall where there is hardly any particle since particles begin to fall as soon as it enters the separator due to sudden reduction in velocity because of large cross-section of the separator (Fig. 6.11). The trajectory of the solids in the separator corresponds to different velocity at different height (Fig. 6.12).

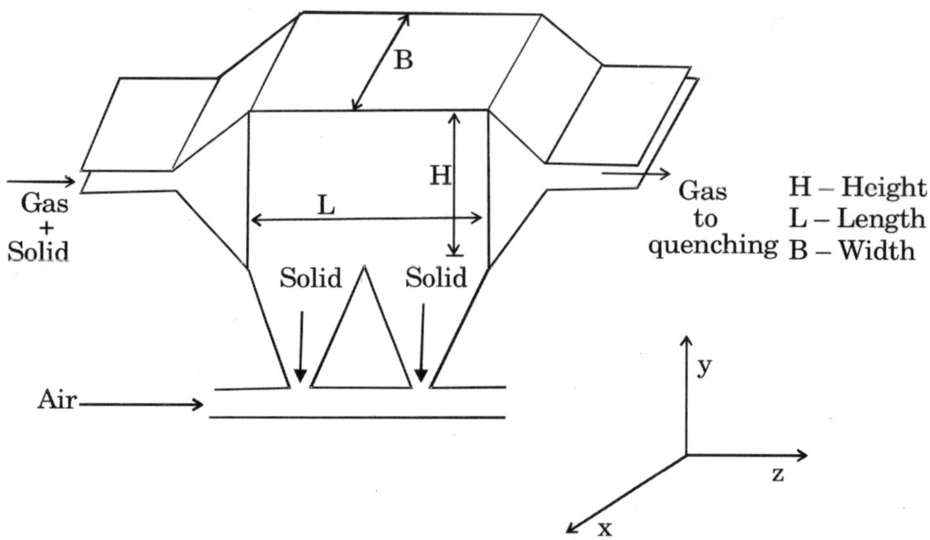

Figure 6.11

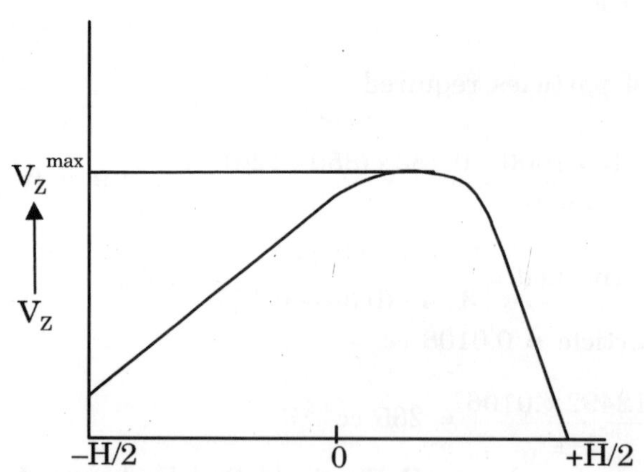

Figure 6.12: Velocity profile in the gas solid separator of vapour naphtha

(*b*) Flow rate of ethylene (F)

$$dF = E(\theta)\, V_z\, B\, dy \qquad ...(27)$$

where E(θ) is concentration of ethylene, a function of contact time in the separator.

$V_z = f(y)$ as is evident from Figure 6.12.

Expression for V_z in terms of y is to be developed from velocity measurements in a typical gravity separator (cold flow experiment) for same size material and same density as used in the preheater and cracker but using a gas which won't undergo any reaction at the conditions of experiment but has similar physical properties as naphtha at 800 to 850°C.

$$\therefore \quad F = \int_{-H/2}^{+H/2} E(\theta) V_z(y)\, B dy = \phi\,(B, H, L) \qquad ...(28)$$

$$\theta = \frac{L}{V_z(y)}$$

now, $$\frac{H}{V_t} = \frac{L}{V_{z_{av}}} \qquad ...(29)$$

where $$V_t = \sqrt{\frac{4g(\rho_p - \rho_{gas})D_p}{3C_D \rho_{gas}}}$$

$$= \sqrt{\frac{4g\rho_p D_p}{3C_D \rho_{gas}}} \qquad \because \quad \rho_p >> \rho_{gas} \qquad ...(30)$$

D_p, ρ_p, ρ_{gas} are known

and $$V_{z_{av}} = \int_{-H/2}^{+H/2} \frac{V_z(y) dy}{H} = \psi(H) \qquad ...(31)$$

From equations (29), (30) and (31) one can get L = ε(H) ...(32)

Substituting this in equation (28) one gets

$F = \delta\,(B, H)$ ...(33)

Q (gas flow rate) Kg/hr

= Flow rate of naphtha

$$= B \frac{\int_{-H/2}^{+H/2} \int_{O}^{L} V_z(y) \rho(y,z) dy dz}{L} = G\,(B, H, L) \qquad ...(34)$$

$$= G'\,(B, H) \qquad ...(35)$$

$\because \quad L = \varepsilon\,(H)$

Density of the gas ρ in the separator is a function of both y and z, to be obtained by measurement.

∵ Q is known

from (35) and (33) $F = D\,(H) = I(B)$ ...(36)

For maximum F, $\dfrac{dF}{dH} = 0$ ...(37)

Therefore, one gets values of 'H', 'B' and 'L' from equations (37), (36) and (32).

6.7 VERIFICATION OF THE QUICK HEATING CONCEPT FOLLOWED BY CRACKING OF NAPHTHA VAPOUR

Preliminary experiments have been carried out for preheating followed by cracking of naphtha vapour and it has been found that the concept works for different preheater and cracker designs using different shapes of particle described in the text. Amongst the various preheater and cracker designs Moving Bed Preheater Design has been found to be least efficient due to the fact that it is limited by the efficiency of heating and nature of velocity profile in gas-solid separator, which gives different yields from cracking reactions (due to different contact time along the height of the separator), which may be poor when considered in totality over the whole cross-section. The quick heating concept described in the text may also be applicable to heating gases other than naphtha vapour.

NOMENCLATURE

A_p	–	surface area of each particle
A_c	–	cross-sectional area of preheater
B	–	width of gas–solid separator
$C_{P_{Naphtha}}$	–	specific heat of the Naphtha vapour
$C_{P_{solids}}$	–	specific heat of the particle
C_D	–	drag coefficient
D_p	–	particle size in meter
$E(\theta)$	–	concentration of ethylene
F	–	flow rate of ethylene
H	–	height of gas-solid separator
G	–	mass velocity of gas
L	–	length of gas-solid separator/length of preheater
N_c	–	number of particles per unit volume of preheater
Q	–	volumetric gas flow rate
r_1	–	inner radius of preheater for cross flow where gas-solid contact occurs
R	–	outer radius of preheater for cross flow from where gas flows into cracker
t_o	–	initial temperature of vapour
t_s	–	temperature of particle
V_z	–	point velocity of vapour
$V_{z_{av}}$	–	average velocity of vapour

V_{p_o}	–	initial velocity of particle
V_{py}	–	velocity of particle at any height y
V_t	–	terminal velocity of particle
$\overline{V_{OM}}$	–	minimum fluidization velocity
W_p	–	weight per particle
W_{Naphtha}	–	Flow rate of Naphtha
W_{solids}	–	Flow rate of particles
y_f	–	wave front thickness
y	–	any length of preheater

Greek letters

ρ_p	–	particle density
ρ_{yv}, ρ_{rv}	–	density of gas
μ_g	–	viscosity of gas
ε_M	–	minimum porosity

REFERENCES

1. Modern Petroleum Technology by B.K. Bhaskara Rao, Oxford and IBH Publishing Company.
2. Unit Operations of Chemical Engineering by McCabe, Smith and Harriott, McGraw Hill 2001.
3. Reaction Kinetics for Chemical Engineers by S.M. Walas, McGraw Hill 1959.
4. Indian Patent Applied, 224/KOL/2006 by Dr. Amalesh Sirkar.
5. Indian Patent Applied, 1319/KOL/2009 by Dr. Amalesh Sirkar.

Section I

CHAPTER 7

Catalyst Recovery and Reuse Technology/Minerals Separation Technology

7.1 CATALYST RECOVERY TECHNOLOGY

In the process of hydrogenation of crude oil using Molybdenum catalyst, more commonly known as the H–oil process developed by Hydrocarbon Research Inc. (New Jersey, USA) demetalisation of the crude *i.e.,* removal of vanadium and nickel also takes place.

H–oil process is usually carried out in an ebullated bed reactor where during the operation due to attrition lots of fines of different sizes of catalyst particles are produced. Also, the different size catalyst particles contain different amount of vanadium and nickel.

The objective is to recover those catalyst particles, which contain less vanadium and nickel so that these can be reused after burning the coke and sulfur containing compounds. Depending on the amount of recoverable catalyst fresh catalyst make up is determined.

Now the question is how to segregate the catalyst particles. One answer may be liquid fluidization. Those who have studied fluidization should come up with this suggestion easily. At HRI we carried out fluidized bed segregation using water. Note that it is the difference in size and density which factors are responsible for segregation in a liquid fluidized bed. As a result, catalyst particles which are smaller in size and which contain less metals (lower density) concentrate towards the upper region of fluidized bed and the bigger and larger metal content catalyst particles concentrate near the bottom region. This can be shown mathematically as follows:

The segregation forces are weight of the particles acting downwards, the buoyancy force acting upwards and the drag force due to fluid motion also acting in the upward direction.

Mathematically this can be represented as follows,[1] considering each particle to be like a sphere:

$$\frac{4}{3}\pi R^3 \rho_p g = \frac{4}{3}\pi R^3 \rho_f g + 6\pi\mu_f u_t R \qquad ...(1)$$

where R is the radius of the sphere, ρ_p is the density of the particle, ρ_f is the density of the fluid, μ_f is fluid viscosity and u_t is the "terminal velocity".

$$\therefore \quad u_t = \frac{4R^2(\rho_p - \rho_f)g}{18\mu_f} \qquad ...(2)$$

The height of expanded bed can be related to particle density, bulk density and other parameters on the basis of Richardson-Zaki correlation for particulate fluidization as follows:

Richardson-Zaki correlation:

$$\varepsilon = \left(\frac{u_f}{u_{t_n}}\right)^{1/m} \qquad ...(3)$$

where ε is bed voidage, u_f is liquid velocity, u_{tn} is terminal velocity. 'm' takes up different values depending on the value of Reynold's No. based on terminal velocity and ratio of particle size and tube size.

1– ε = Volume of Particles/Expanded Packed Volume

where ε is fluidized bed voidage

$$h_n = \frac{4/3\pi R_n^3 N_n}{\frac{\pi}{4} d_s^2 \left(1 - \left(\frac{u_f}{u_{t_n}}\right)^{1/m}\right)} \qquad ...(4)$$

Where h_n is height of expanded bed, N_n is number of particles of size R_n and density ρ_{pn} u_{tn} is terminal velocity of the particles of size R_n and density ρ_{pn}, d_s is diameter of segregator.

If ε_{fn} is fixed bed voidage and h_{fn} is fixed bed height, then

$$1 - \varepsilon_{fn} = \frac{4/3\pi R_n^3 N_n}{\pi/4 d_s^2 h_{fn}} \qquad ...(5)$$

But $\quad 1 - \varepsilon_{fn} = \dfrac{\rho_{bn}}{\rho_{pn}}$

where ρ_{bn} is bulk density

$\therefore \quad N_n \times 4/3\ \Pi\ R_n^3 = \Pi/4\ d_s^2\ h_{fn} \times \rho_{bn}/\rho_{pn}$

Substituting this in equation (4)

$$h_n = \frac{h_{fn} p_{bn}}{\rho_{pn}(1 - (u_f/u_{tn})^{1/m})} \qquad ...(6)$$

It is apparent from above equation that for a given liquid velocity, higher the density of the particles and bigger the size lower is height of the expanded bed, which was stated earlier and now shown mathematically. Lower height means heavier and bigger size particles will concentrate near the bottom region.

Example: 90 gms of used catalyst particles obtained from H-oil reactor were fluidized by water at 70°F. The segregator diameter was 0.76 inch and the fixed bed height was 12 inches.

Total catalyst sample was analyzed for particle size. Water velocity used was 0.11 fps and it was found that at this velocity expanded bed was twice the height of fixed bed *i.e.,* 100% bed expansion. The segregator height was $2\frac{1'}{2}$. The results of segregation experiment are given in Table 7.1.

The bed was analyzed for metal content from top to bottom as shown in Table 7.1. It is obvious from the table that lower metal content particles having less density existed in the upper region whereas higher metal content ones concentrated near the bottom region. It was also found from analysis of metal content that – 20 mesh particles in the top region contained less metal than the ones near the bottom region. This is obvious because particle density is directly proportional to metal content.

Table 7.1

Sl. no.	*Layer no. from top*	*(Ni + Va)/Mo*	*Particle size*
1.	Top layer	1.06	+ 20 mesh
1(*A*)	”	2.47	– 20 mesh
2.	Second layer	1.53	+ 20 mesh
3.	Third layer	1.87	”
4.	Fourth layer	1.89	”
5.	Fifth layer	2.19	”
5(*A*)	”	4.15	– 20 mesh
6.	Sixth layer	2.57	+ 20 mesh
7.	Seventh layer	2.66	”
8.	Eighth layer	2.96	”
9.	Nineth layer	3.55	”
10.	Tenth layer	3.67	”
10(*A*)	”	5.59	– 20 mesh

Note : All layers are of equal height

The next question is how to recover portion of the bed of catalysts for reuse. The recovery can be done either from top or from side of the segregator. There are two ways to recover from top – (1) by increasing the fluidizing medium velocity so that the bed will expand and will overflow from the top or (2) by pulling vacuum after draining the bed and drying.

The disadvantage of the first method is that with increase in bed expansion beyond an optimum value, segregation is destroyed due to mixing currents and hence the material which overflows is a mixture of high and low density particles. This is not so in method 2, where material flows like pneumatic conveying from the top of the vessel of segregated bed which can be controlled by breaking the vacuum whenever desired.

However, method 1 is simpler because it does not require many pieces of equipment and simple to operate than method 2 (described later in this chapter).

Commercial segregator designs for method 1 are shown in Figs. 7.1 and 7.2 with a description of the mode of operation. In Fig. 7.3, similarly a system design on the basis of the principle of collection of segregated bed by method 2 is shown with description of the system.

7.2 RECOVERY FROM SIDE

1. This can be done by means of sliding doors on slits or opening at different levels of the segregator. The openings are rectangular in shape along the side of the cylindrical segregator. Segregator materials can be scooped out through the openings. This method is superior to method 1 described above but not superior to method 2 because in method 2, by controlling the vacuum level and breaking it at short intervals differential (small) amounts of segregated bed can be recovered whereas in the side recovery there exists a minimum height of layers below which it is not possible to recover the bed. Apart from this, in side recovery, there could be a problem of leakage of the fluidizing medium during segregation through the sliding doors. These are some of the disadvantages of side recovery compared to method 1.
2. Side withdrawal can also be done using the vacuum technique through small openings on the side of the segregator at different levels using one at a time. This way, it is easier to withdraw layers by stopping the vacuum pump when the level of solids reaches the opening height if we use a transparent segregator through which we can see the solids.
3. Side withdrawal can also be done while the bed is in fluidized condition through small openings on the side, using one opening at a time.

Method 1 (Refer Figs. 7.1 and 7.2)

In laboratory, very sharp separation between lighter and heavier catalyst particles was achieved using a small quantity of spent catalyst. However, in a commercial size segregator with significantly large solid loading, sharp separation may not be achieved in one segregation step due to lateral mixing currents. To improve the segregation in a commercial segregator therefore, two techniques can be used as follows:

1. If we use a shallower segregator than the lab scale then we have to use more than one stage of segregation[(5)] (Fig. 7.1).

After initial segregation, separate the bed in two portions (50/50). Followed by that resegregate the two portions. Then take 80 to 90% of the resegregated upper half and 10 to 20% of the resegregated lower half. By doing this we would improve the overall segregation efficiency of recovering most of the lighter catalyst particles. The design of the vessel to do more than one stage of segregation is shown in Fig. 7.1.

At first, segregation will be carried out by circulating fluidizing medium through the outermost annulus. Then, upper half of the segregated bed will be transferred into the next annulus by increasing the fluidizing medium velocity. Followed by that both the upper half and the lower half will be resegregated. The fluidizing medium will be flowing through both the outermost and the next inner annulus. Followed by resegregation desired percentage of resegregated upper half will be transferred to the innermost cylinder and recovered while rest of the resegregated upper half will be thrown out. Next desired percentage of resegregated lower half will be transferred from outermost annulus to the next one and eventually recovered while rest of the resegregated lower half of the bed will be discarded.

2. We can use a multitubular segregator having each tube L/D same as that of the lab scale. This technique of segregator design is better than method (1) mentioned above (Fig. 7.1).

The segregator is like a tubular heat exchanger with each tube having same L/D ratio as the lab scale one. The tubes stand straight by means of tie rods. The unsegregated materials are first dumped inside the tubes. Next water is injected inside the tubes from bottom and withdrawn from the top of the segregator. The water will be collected in the space between the tubes and withdrawn from the bottom through a common header pipe. After desired segregation is achieved the recovery of layers can be done by increasing velocity of water which will push the beds in upward direction inside each tube and water along with particles will be collected in the annular space, and taken out of the segregator through the header pipe into another vessel from which water can be drained out and solids are recovered.

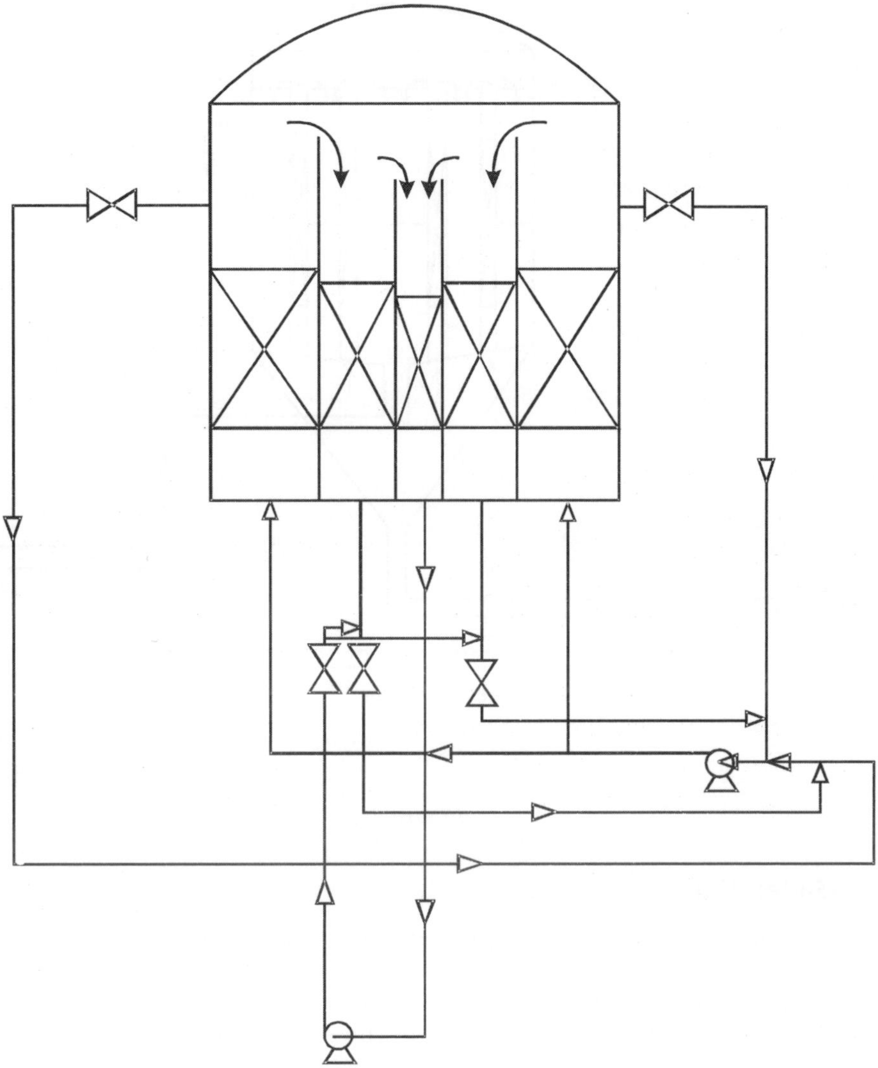

Figure 7.1

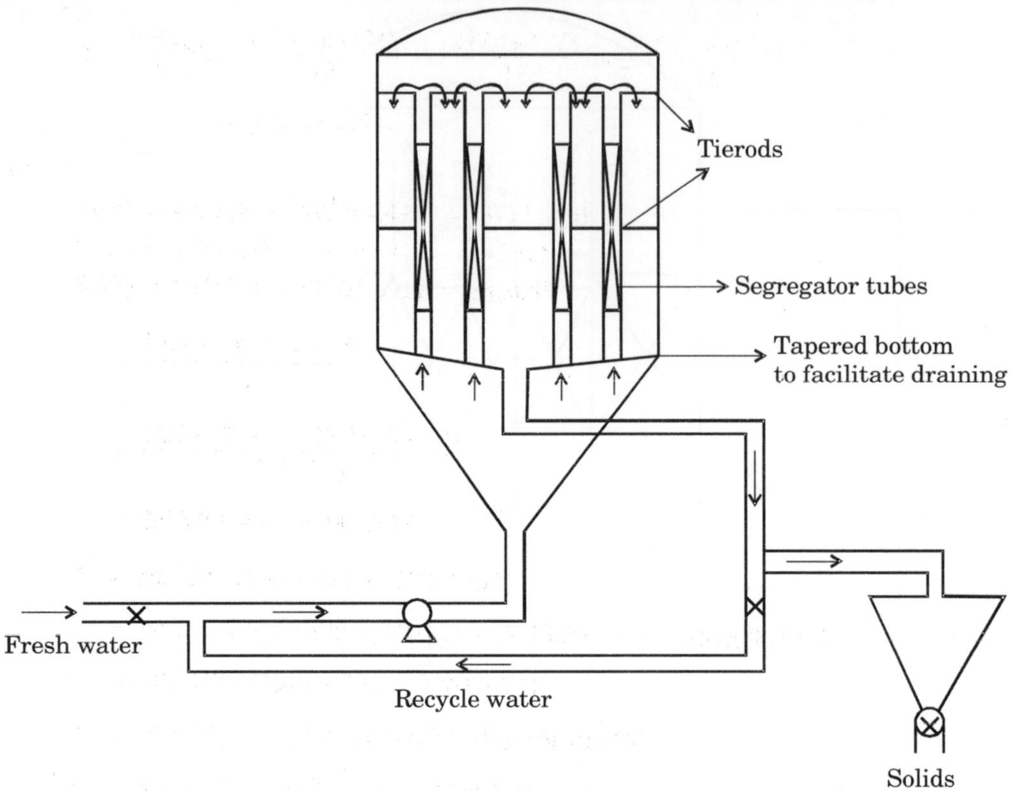

Figure 7.2

Method 2 (Refer Fig. 7.3)

The segregator is connected to a hopper type vessel with a valve at the bottom and another at the top. Also, a bypass valve is located on the suction side of the vacuum pump for controlling vacuum level. Soon after we start the vacuum pump the solids will start coming into the hopper and the vacuum level in the segregator will start increasing. The rate of solids removal can be controlled by operating the bypass valve *i.e.,* controlling the vacuum. After desired time period when differential amount of solids are transferred shut off the pump and open the valve at the top of the hopper to quickly bring the system to atmospheric pressure. Next, the valve at the bottom of the hopper is opened and solids are collected. After solids removal one should close both the valves of the hopper and start evacuating the system again and collect small quantity (desired) of additional solids and repeat the procedure as described above.

At the top of the hopper, a perforated plate is to be installed at the outlet to prevent solids from flowing to the pump suction when vacuum is on. Also, to prevent solids from going back to the segregator from the hopper when the vacuum is broken (by opening the valve at the top of the hopper) a valve is installed in the outlet of the segregator which is kept closed. This valve is kept open when vacuum is on, so that solids flow from segregator to hopper.

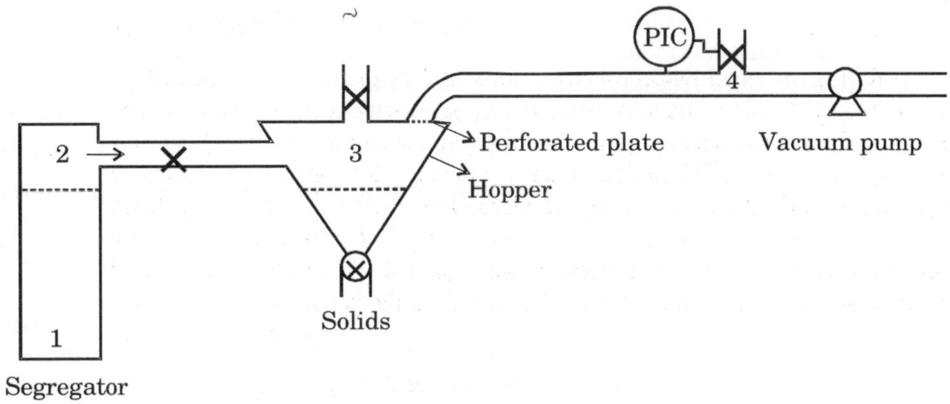

Figure 7.3

7.3 THEORY BEHIND VACUUM WITHDRAWAL (FIG. 7.3)

Due to creation of vacuum, the solid particles will be lifted and pulled from the segregated bed. The energy required to pull the solids is given by the product of total pressure drop (for flow of solids and air) and volumetric flow rate of air and solids combined. The total pressure drop is given by the following equation:

$$\Delta P_{Total} = \Delta P_{lift} + \Delta P_{vertical} + \Delta P_{bend} + \Delta P_{valve} + \Delta P_{horizontal} \quad ...(7)$$

$$\Delta P_{lift} = \frac{W_s'}{\frac{\pi}{4} d^2_s} \quad ...(8)$$

$$\Delta P_{vertical} = \frac{V_{G_1}{}^2 \rho_g}{2g_c} + \frac{W_{s_1} V_p}{g_c} + \frac{2L_v}{g_c d_s}(f_g \rho_g V_{G_1}{}^2 + f_s W_{s_1} V_p) + \frac{W_{s_1} L_v}{V_p} \quad ...(9)$$

$$\Delta P_{bend} = 2f_b \rho_g V_{G_2}{}^2 \quad ...(10)$$

$$\Delta P_{valve} = \frac{600 f_g V_{G_2}{}^2}{g_c} \quad ...(11)$$

$$\Delta P_{horizontal} = \frac{V_{G_2}{}^2 \rho_g}{2g_c} + \frac{W_{S_2} V_p}{g_c} + \frac{2L_h}{g_c d_t}(f_g \rho_g V_{G_2}{}^2 + f_s W_{s_2} V_p) \quad ...(12)$$

$$f_s = \frac{0.05}{V_p} \quad ...(13)$$

The above expressions for pressure drop are taken from references 2 and 3.

V_{G_1} and V_{G_2} (air velocity in segregator and in the pipe connecting segregator and hopper) are obtained from W, the mass flow rate of leakage air (in lb/hr) which is calculated using the following expression[(4)]:

$$W = cV^{2/3} \quad ...(14)$$

Where c is a constant = 0.2 when $p < 90$ torr and 0.08 when $3 > p < 20$, and V is total volume of the vacuum system in ft^3. Hopper volume is based on 50% (maximum) removal of segregated bed with 30% free space above the solids. W_s', the total weight of solid particles in the segregator was 90 gm. Segregator to hopper connecting pipe's diameter is 1/4″ and length 1 ft. Porosity of settled bed is 0.36.

For a given vacuum level the total pressure drop equation is solved for particle velocity V_p. Since, the pressure drop equation is quadratic in nature in V_p, two values have been obtained for each vacuum level for a given withdrawal rate of solids and the realistic value of V_p has been taken and plotted in Figs. 7.4, 7.5 and 7.6 for 3 different vacuum levels. Note that the V_p calculated is the average velocity of different size and different density particles. ΔP_{lift} will actually vary with time during the withdrawal, becoming less due to less weight of solids per unit area of bed as time progresses.

Results and Discussions

From Figures 7.4, 7.5 and 7.6 it is clear that as vacuum level is increased particle velocities increase for a given withdrawal rate of solids, the reason being the fact that higher vacuum will suck the particles quicker from the bed. Also, for a given vacuum level particle velocities obtained by calculation varies from 0.0001 to 0.008 m/sec and the range of solids withdrawal rate for this velocity range is 1 to 8 Kg/m^2sec. The time of withdrawal has been varied from 4 to 160 seconds.

Conclusion

Theoretical calculations have been carried out using pressure drop equation for withdrawal of solids from the segregated bed using vacuum technique. Experiments should be carried out using the vacuum system proposed to find out what is the true particle velocity and solids withdrawal rate for a given vacuum level.

7.4 MINERALS SEPARATION TECHNOLOGY

When two or more mineral matters occur together in nature, particulate fluidization using water or a suitable liquid can be used to segregate the individual mineral matter using the same principle as described earlier in this chapter. For example, if we want to recover some of the mineral matters from coal ash *e.g.*, silica, alumina which are cracking catalysts then we can adopt method 1 (Fig. 7.2) described earlier and method 2 which are superior to all other methods in terms of segregation and efficiency of separation of segregated bed. Note that silica and alumina densities are different.

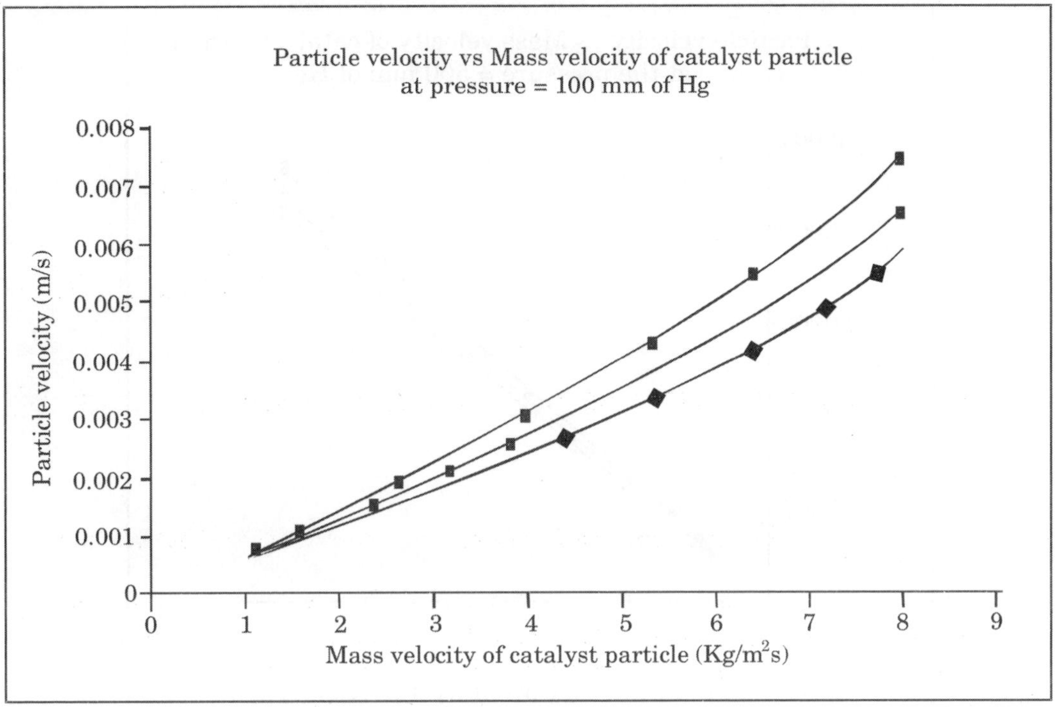

Figure 7.4

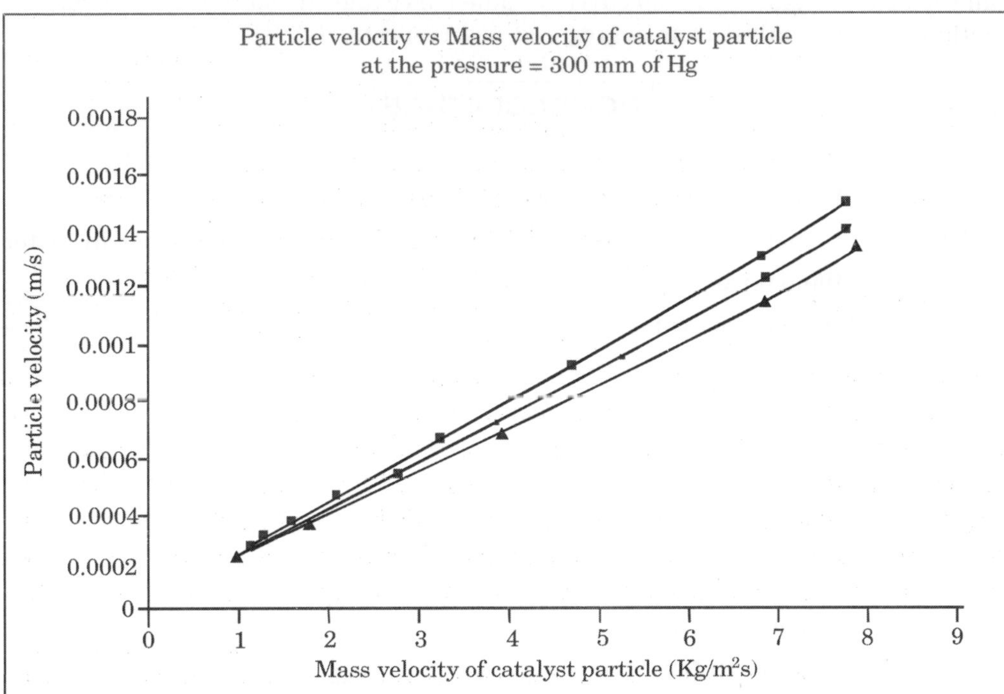

Figure 7.5

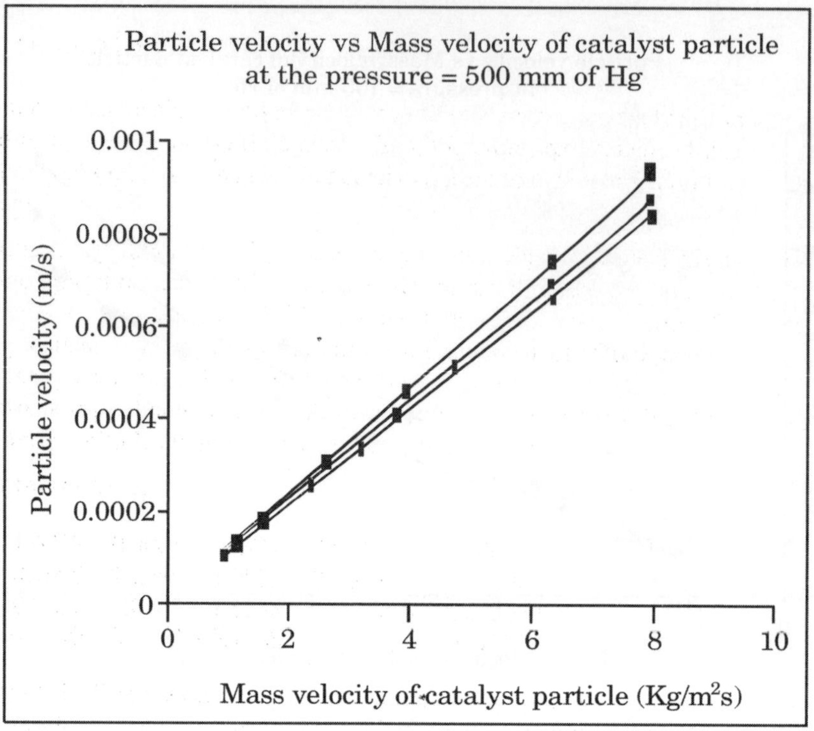

Figure 7.6

In all the above 3 figures the top curve, middle curve and the bottom one are for 50%, 30% & 10% withdrawal of catalyst from segregated bed respectively.

NOMENCLATURE

c	–	constant
d_s	–	diameter of segregated bed in meter
d_t	–	diameter of pipe connecting the segregator and the hopper in meter
d_p	–	diameter of a spherical particle in meter
f_b	–	Friction factor for flow through bend
f_g	–	Fanning friction factor
f_s	–	particle to wall friction factor
g_c	–	gravitational constant
L_v	–	length of vertical section in meter
L_h	–	length of horizontal section in meter
Mo	–	Molybdenum
Ni	–	Nickel
p	–	pressure in torr
ΔP_{lift}	–	pressure drop due to lifting of particles in Kg force/m^2

$\Delta P_{vertical}$ – pressure drop in the vertical section (segregator) in Kg force/m^2

ΔP_{bend} – pressure drop in the bend between vertical and horizontal section in Kg force/m^2

$\Delta P_{horizontal}$ – pressure drop in the horizontal section before the hopper in Kg force/m^2

ΔP_{valve} – pressure drop in the valve in the line connecting the segregator and the hopper in Kg force/m^2

V – total volume of vacuum system in ft^3

Va – Vanadium

V_{G_1} – leakage air velocity in vertical section in m/sec

V_{G_2} – leakage air velocity in horizontal section in m/sec

V_p – particle velocity in m/sec

W – mass flow rate of leakage air in lb/hr

W_{s_1} – mass velocity of particles in vertical section in Kg/m^2 sec

W_{s_2} – mass velocity of particles in horizontal section in Kg/m^2 sec

W_s' – total weight of solid particles in segregator in Kg

Greek letters : ρ_g – air density in Kg/m^3

ρ_p – particle density in Kg/m^3

REFERENCES

1. 'Transport Phenomena' by Bird, Stewart and Lightfoot, John Wiley and Sons (1960).
2. F.A. Zenz and D.F. Othmer, 'Fluidization and Fluid Particle System', First edition, Reynold Publishing Corporation, New York, 1960.
3. D. Kunii and O. Levenspiel, 'Fluidization Engineering', Butterworth-Heinmann Publishing, 2nd edition, 1991.
4. S.M. Walas, 'Rules of Thumb', Chemical Engineering, March 16, 1987.
5. US Patent 4,287,088 (1981) by Dr. Amalesh Sirkar.
6. Indian Patent 177075 (1997) by Dr. Amalesh Sirkar.

Section I

CHAPTER 8

Hydrogenation of Vegetable Oils and Fatty Acids

It is well known to hydrogenate vegetable oils and fatty acids using nickel catalyst. Attempts have been made to carry out hydrogenation(1) at around 180˚C. At this temperature, the overall rate of hydrogenation is controlled by the rate of diffusion of hydrogen to the catalyst surface, rate of diffusion of oil/fatty acid to the catalyst and diffusion of product from catalyst surface to the bulk of the liquid. It has also been attempted to enhance hydrogenation by using higher stirrer speeds, but this results in leakage of the contents and higher power consumption. Besides, at high stirrer speed there could be vortex formation resulting in non-uniform hydrogenation. Several researchers have worked on different routes and have suggested various techniques as follows:

1. In an old patent GB 657188 particulate nickel is suspended with or without promoters and the medium is agitated mechanically. It suffers from the same drawback mentioned above.
2. In another patent GB 1154230 the starting method conditions the feed which contains free sterol which are substantially gum free. Conventional nickel catalysts are used. The overall production rate is not very high compared to the inputs of materials, time taken and energy consumed.

In the process suggested here several alternatives are recommended to improve the rates of mass transfer.

8.1 A NEW PROCESS

To tackle the conventional problems attempts have been made from several angles as follows:

1. To vibrate the reactor mechanically at selected and regulated speeds based on the contents of the reactor.
2. To employ specific shape of the nickel catalyst thereby dispensing with the old concept of using particulate catalysts. It has been found that rods of nickel bent to form a cage

like construction, most preferably a sphere does the job better (Figure 8.1). The rods are circular in cross-section.

The mechanism of improved mass transfer rate is as follows:

Due to motion of the liquid past the rods, fats produced at the surface of the catalyst are swept away at a rate faster than molecular diffusion. Although in a stirred system there occurs same sweeping effect the power consumption is less(2) in the present process of vibratory motion and hence cost of hydrogenation is less. Power utilization is more effective here than stirrer.

Time required for hydrogenation is much less compared to the conventional process due to higher rate of hydrogenation. Hence, production capacity is increased. Another advantage of the suggested process is the fact that it is easier to separate the large shaped catalyst particles from the hydrogenated product compared to the conventional process where fine nickel particles are used and some of which become finer during operation due to stirrer hitting the particles.

8.2 VERIFICATION OF THE NEW PROCESS OF HYDROGENATION

Some preliminary experiments have been carried out and it has been found that for same degree of hydrogenation the new process consumes much less power and takes less time than the conventional process, besides the fact that leakage problem is completely eliminated.

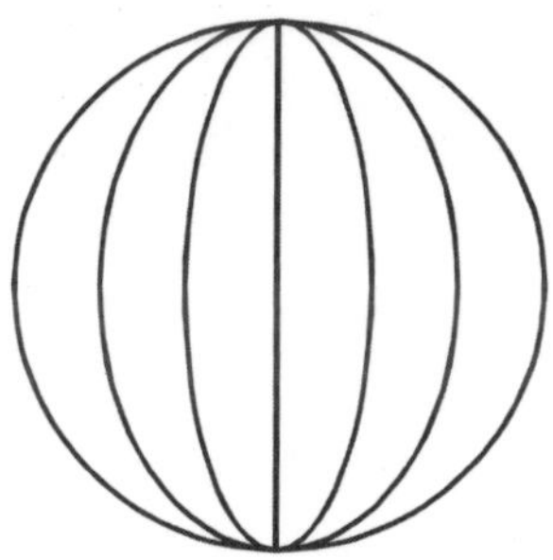

Figure 8.1

REFERENCES

1. Dryden's outlines of Chemical Technology edited by M. Gopala Rao & Marshall Sittig, 2nd edition (East West Press), 1973.
2. Indian Patent Applied in 2008, 825/KOL/2007 by Dr. Amalesh Sirkar.

SECTION II

EQUIPMENT DESIGN AND CREATIVITY

Section II

CHAPTER 1

Mass Transfer Equipment Design

In this chapter design of gas-liquid absorption column, liquid-liquid and solid-liquid extractors have been discussed with emphasis on the novelty associated with their design for improving the efficiency of these equipments.

A novel gas absorption column design concept has been discussed which consists of the use of rocking arrangement and the use of novel packing geometry for improving the efficiency of absorption.

The most common countercurrent liquid-liquid extraction column is a spray column without much of an internal. However, as the theory behind the design of such columns is discussed for transfer of surface-active solutes, the need for internals arises for increasing extraction efficiency.

For handling solid-liquid systems, which are viscous, special design of stirrer is necessary for improving extraction efficiency. This has been discussed later in this chapter.

1.1 ABSORPTION COLUMN

It has been discovered after detailed study of gas-liquid mass transfer in packed columns[14] that giving a rocking motion to the tower in lateral direction the gas and liquid phase resistances to mass transfer can be reduced. The rocking motion generates motion in the gas and liquid film surrounding the packing material and this enhances the rates of mass transfer through the gas and liquid films due to reduction in film thickness. Actually, lateral motion in the liquid film surrounding the packing material causes gas-liquid interface to move which in turn makes the gas film move back and forth. This results in increased diffusive transport of solute from bulk to interfacial region on the gas side and from interface to bulk on liquid side due to reduction in gas and liquid film thickness. Normally, in absence of vibration, both the gas and liquid films are stagnant. Thus, with vibratory motion, for a given absorption equipment the tower volume can be reduced.

Another advantage of lateral movement of the absorption column is elimination of channeling due to freeing the passage of gas and liquid unlike a simple packed column where channeling is a common problem.

The lateral motion is generated by rocking the column as shown in Fig. 1.1.

The column is freely supported on a ring installed near the middle section. The ring is fixed to ground by means of connecting rods. Lateral motion is generated by means of crank slider arrangement (driven by a motor) connected to the lower end of the absorption column.

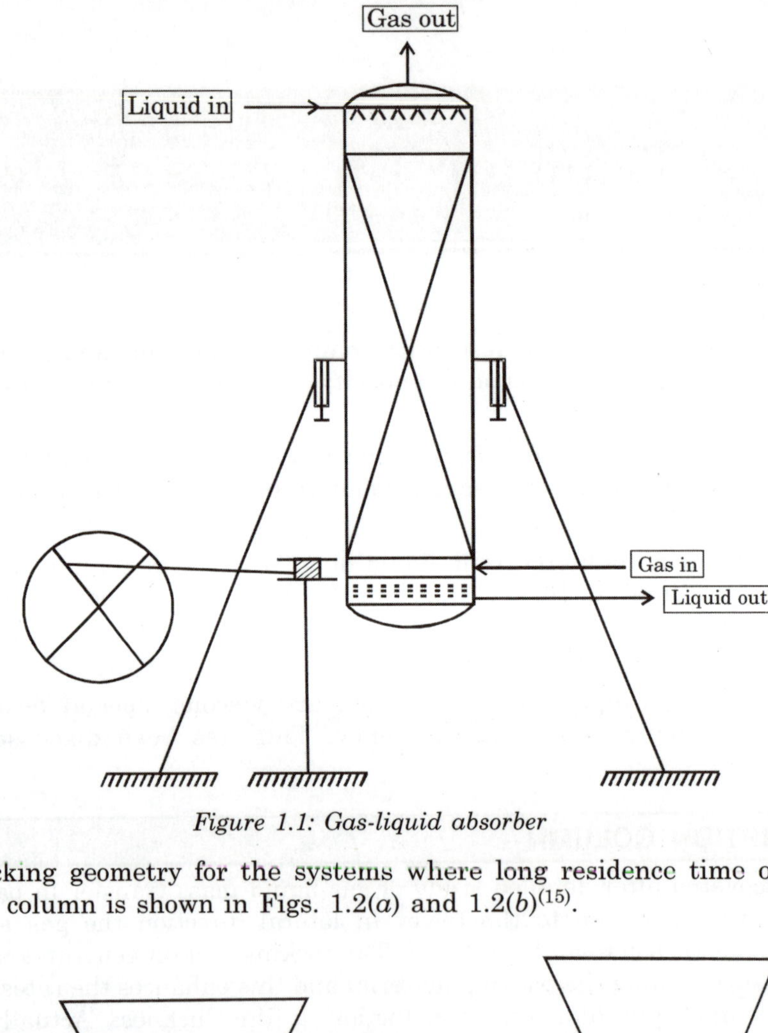

Figure 1.1: Gas-liquid absorber

A novel packing geometry for the systems where long residence time of the liquid is necessary in the column is shown in Figs. 1.2(*a*) and 1.2(*b*)[15].

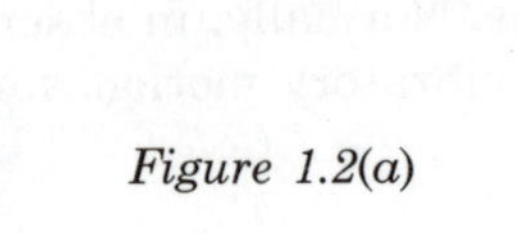

Figure 1.2(a)

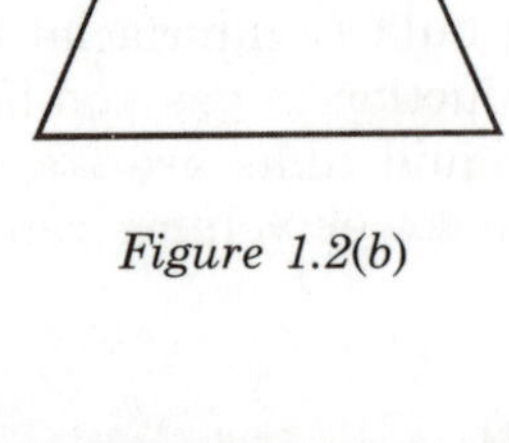

Figure 1.2(b)

The advantage of trough shaped packing is because of the hold up given by the packing for reaction systems. An example of the system is bacterial oxidation of ethanol to acetic acid. Bacterial oxidation will be enhanced if in addition to the trough shaped packing the column is vibrated. This is because the liquid in the trough will be in mixed state. Thus, each packing acts like a stirred tank reactor.

1.2 THEORY OF LIQUID–LIQUID EXTRACTION OF SURFACE ACTIVE SOLUTES

The traditional theory of liquid-liquid extraction does not assume occurrence of any interfacial instability which has been observed by many[1] researchers. Interfacial instability is either of the two types namely, Marangoni Instability[2] and Energy Instability[3]. Marangoni instability causes surface tension driven flow in the vicinity of the interface, which increases the overall mass transfer rate by 2 to 3 times the diffusional rate. Energy instability occurs in the form of ejection of mass from one phase to the other[4] if specific energy of driving force of mass transfer exceeds the free energy of the interface. Due to lower energy barrier of the interface instead of molecular diffusion, mass of solute penetrates via bulk flow from interfacial region of one phase to another through holes produced at the interface. Holes are produced at the interface by the solutes penetrating force due to higher energy level of the molecules. If interfacial tension would have been zero meaning zero barrier, then no interface would exist and solute would have passed easily. But, because there is a physical presence of interface some resistance is offered by the interface and hence there are holes produced for easy passage of solute. The interfacial mass transfer coefficient in such a case is nothing but the velocity with which the molecules pass through the holes from one phase to another. With time as equilibrium is approached the interfacial tension rises and the velocity of molecules decreases due to greater resistance offered by the interface, and hence it has been observed that depth of penetration into the other phase decreases. Marangoni instability induced interfacial turbulence usually occurs when transfer takes place from aqueous to organic direction, whereas ejection of mass occurs in both directions of transfer[5]. However, usually it occurs in organic to aqueous direction of transfer[4]. Energy instability causes increase in overall mass transfer rate much more than due to Marangoni instability. Here, we shall concentrate on design of liquid-liquid extraction column in presence of Energy instability only.

It is expected that consideration of Energy Instability would predict much less height of the spray column than that evaluated on the basis of standard design procedure because of high mass transfer rate across the interface due to instability. It has been shown later in this chapter that if we use multiple feeds from top to bottom of the spray column using internals, then height can be reduced significantly. This is because higher solute concentration at the drop surface reduces interfacial tension significantly due to more adsorption at the interface[17] thus enhancing the interfacial mass transfer rate.

$$\Re_i = A' e^{-F'\sigma_1/RT}$$ [5]

Criterion of Energy Instability[3]

$$(1 - C_{2i}/C_{1i}\ \Re_D)\ RT \quad > \quad \sigma_i\ S_o$$

Specific energy of driving force > Free energy of the interface of mass transfer from phase 1 to phase 2.

Another aspect of interphase mass transfer which has not been considered previously is occurrence of convection within the phase due to unstable density gradient[6,7,8]. Such convection enhances the mass transfer from bulk to interface or from interface to bulk as the case may be[8].

When a solute gets transferred from one phase to the other, unstable density gradient may occur if ρ solute > ρ phase and the phase is lighter than the phase in which transfer occurs. If ρ solute is greater than the phase in which it is transferred and it is the heavy phase, then also density convection may occur due to unstable density gradient. On the other hand if ρ solute < ρ phase and the phase is heavier then also transfer of solute out of this phase may cause density convection. Also, if ρ solute is less than the phase in which it is transferred and the phase is lighter, then also density convection may occur.

Mass transfer coefficient for the phase in which density convection occurs (normally organic phase) can be evaluated using the following equation[8] :

$$\Re_{T_{org}} \, h / D_M = c (Sc)^a (Ra)^b$$

where $c = 0.7$

$a = -1.04$

$b = 0.6$

1.3 DESIGN EQUATIONS

A. Conventional Spray Column (Fig. 1.3)

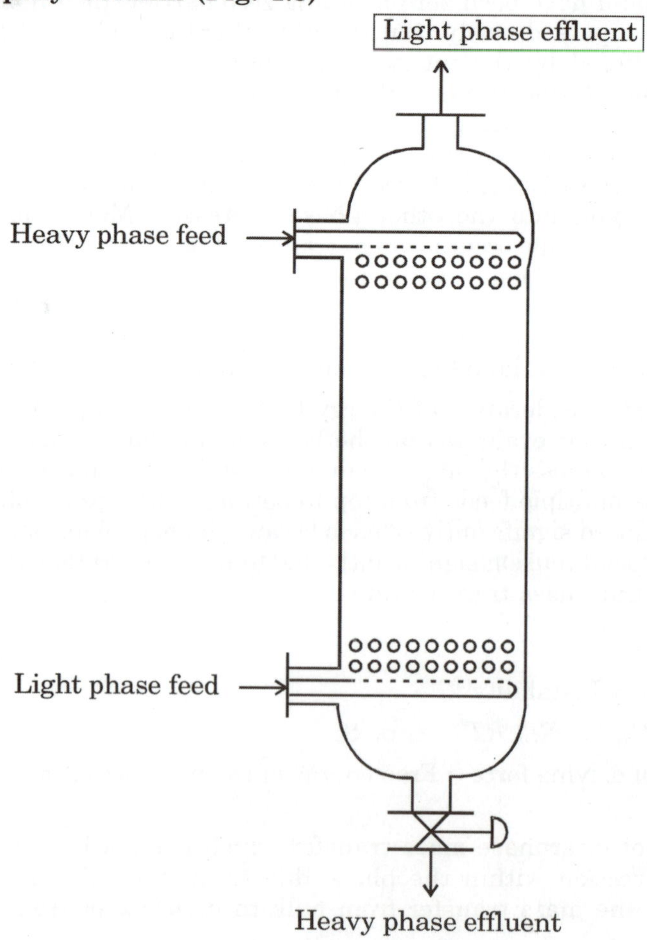

Figure 1.3: Conventional spray column

Assumptions

1. Organic to aqueous phase direction of transfer has been considered because energy instability normally occurs in this direction. However, there are examples where energy instability also occurs in the reverse direction[3].
2. Aqueous phase is dispersed because in aqueous phase interfacial turbulence (ejections) is restricted to a small region close to the interface only. Hence, if drops are small in size there will exist only one concentration due to turbulence within the drop and bulk resistance will be nonexistent.
3. Organic phase is continuous because there occurs density convection in this phase thus, bulk to interface transfer resistance is minimized.
4. No solute is present in the aqueous phase entering the column.
5. Drop size is 1 mm because it has been found that ejections penetrate aqueous phase up to a distance of 2 mm[4] at the beginning of extraction in stagnant system and this distance decreases as one approaches equilibrium.

1. *Mass Balance Equations:*

$$L\left(C^{0}_{org} - C^{top}_{org_{av}}\right) = H_{aq}\, C^{0}_{aq_i} \qquad \text{...(1)}$$

This is overall material balance for the solute in the column.

For solute balance between bottom and nth section up the column the following equation holds good:

$$L\left(C^{0}_{org} - C^{n}_{org_{av}}\right) = H_{aq}\left(C^{0}_{aq_i} - C^{n}_{aq_i}\right) \qquad \text{...(2)}$$

2. *Rate Equations:*

$$H_{aq}\, dC^{n}_{aq_i} = \Re^{n}_{i_{av}}\left(C^{n}_{org_i} - \frac{C^{n}_{aq_i}}{\Re_D}\right)_{av} A_{drop}\, dv_n \qquad \text{...(3)}$$

Also, $$Lz_f\left(-dC^{n}_{org_i}\right)A_{drop} = \Re^{n}_{i_{av}}\left(C^{n}_{org_i} - \frac{C^{n}_{aq_i}}{\Re_D}\right)_{av} A_{drop}\, dv_n \qquad \text{...(4)}$$

and $$Lz_f\left(-dC^{n}_{org_i}\right)A_{drop} = \Re^{n}_{T_{org_{av}}}\left(C^{n}_{org_{av}} - C^{n}_{org_i}\right)_{av} A_{drop}\, dv_n \qquad \text{...(5)}$$

Note that the concentration gradients in equation 3, 4 and 5 are arithmetic average of the gradients between nth and (n – 1)th stage.

3. *Mass Balance for solute between Bottom section of the column and End of First volume Increment where Instability stops:*

$$L\left(C^{0}_{org} - C^{1}_{org_{av}}\right) = H_{aq}\left(C^{0}_{aq_i} - C^{1}_{aq_i}\right) \qquad \text{...(6)}$$

4. *Mathematical Criterion where Instability Stops[3]:*

$$\frac{M}{H} = 0.99 = \frac{RT\left[1 - \frac{C^1_{aq_i}}{C^1_{org_i}\,\Re_D}\right]}{\sigma^1_i S_o} \quad \text{...(7)}$$

where $$\sigma^1_i = \sigma_{eq} - A + B \log \frac{C^1_{aq_i}{}^{(9)}}{C^1_{org_i}} \quad \text{...(8)}$$

where σ_{eq}, A and B and are constants for a ternary system.

5. *Interfacial Mass Transfer Coefficient Correlation*[5]:

$$\text{Log } \Re^n_i = a - bC^0 - F'\sigma^n_i \quad \text{...(9)}$$

where a, b, and F' are constants for the specific ternary system.

6. *Bulk Organic Phase Mass Transfer Coefficient Correlation*[8]:

$$\frac{\Re_{T_{org}} h_{av}}{D_M} = 0.7\ (Sc)^{-1.04}\ (Ra)^{0.6} \quad \text{...(10)}$$

where $$Sc = \frac{\mu_{org}}{\rho_{org} D_M} \text{ and } Ra = \frac{\Delta\rho_{org} h^3_{av} g}{\mu_{org} D_M}$$

where $h = \frac{dv}{A_c}$ and $h_{av} = \frac{h}{2}$, since density eddies would rise from different heights within dv.

For the region where there exists no instability since $\frac{M}{H} < 1.0$, the following equations will be applicable to calculate dv. At the end of this region instability will be restored due to increase in C_{org_i} and hence lowering of σ_i as will be evident from equations (13) and (16) given below. For the 1st stage of no instability following equations hold good:

7. *Mass Balance Equation:*

$$L\left(C^1_{org_{av}} - C^2_{org_{av}}\right) = H_{aq}\left(C^1_{aq_i} - C^2_{aq_{av}}\right) \quad \text{...(11)}$$

8. *Rate Equations:*

$$H_{aq} dC^2_{aq_{av}} = K^2_{D_{av}}\left(C^2_{org_i} - \frac{C^2_{aq_i}}{\Re_D}\right)_{av} A_{drop}\, dv_2 \quad \text{...(12)}$$

$$\underbrace{Lz_f A_{drop} dC^2_{org_i}} = \Re_{T_{orgav}}\left(C^2_{org_{av}} - C^2_{org_i}\right)_{av} A_{drop} dv_2 - K^2_{D_{av}}\left(C^2_{org_i} - \frac{C^2_{aq_i}}{\Re_D}\right)_{av} A_{drop} dv_2 \quad \text{...(13)}$$

Increase in C_{org_i} in the interfacial region.

Note that the concentration gradients in equations (12) and (13) are arithmetic average of the concentration gradients at the end of 1st stage and end of 2nd stage.

K_D is given by same expression as $\Re_i$ (equation 9) but, having different values of a, b and F'

$\Re_{T_{org}}$ is also given by equation 10

Also, $$H_{aq}\,dC_{aq_{av}} = n\left(4\pi r_p^2\right)D_W \left.\frac{\partial C_{aq}}{\partial r}\right|_{r=r_p}$$ ...(14)

and $$n\left(4\pi r_p^2\right) = A_{drop}\,dv$$ ...(15)

9. *Mathematical criterion where instability is restored:*

$$\frac{M}{H} = 1.01 = \frac{RT\left(1 - \dfrac{C^2_{aq_i}}{C^2_{org_i}\,\Re_D}\right)}{\sigma_i^2 S_o}$$ ...(16)

where $$\sigma_i^2 = \sigma_{eq} - A + B\log\frac{C^2_{aq_i}}{C^2_{org_i}}$$ ...(17)

Next, consider again instability section and calculate corresponding dv and 'h' and continue as before up to the top section of the column with alternate regions of instability and no instability. The total height of the column is the summation of 'h' for each volume increment.

Note: $\left.\dfrac{\partial C_{aq}}{\partial r}\right|_{r=r_p}$ can be evaluated by setting up time variant differential mass balance equation for the portion of column where instability is non-existent and solving such an equation.

B. Multiple Feed Spray Column

In conventional spray column there are regions where there occur no instability and hence the mass transfer rates are lower. In order to have instability throughout the column multiple entry point of feed to the column at specific locations could be made to further cut down the height of the column. The rate of injection of fresh feed may vary from bottom to top such that desired increase in concentration of solute at the interface on the organic side occurs to restart instability. The increase in concentration of solute reduces the interfacial tension due to more adsorption at the interface[17], thereby increasing the rate of mass transfer across the interface. Feeding of organic phase containing solute to upper regions of the column instead of all of it at the bottom is against the conventional liquid-liquid extraction. But, here actual mass transfer is more due to lower interfacial tension in presence of higher solute concentration. Actually, more mass gets ejected from interfacial region to inside the drops due to lower interfacial tension in comparison to the amount of fresh mass injected in the organic phase through multiple feeding. The following design equations determine the height of such a column:

Design Equations

Mass Balance Equations:

The overall mass balance for the column is

$$\left(LC^0_{org} + R_1 C^0_{org} + R_2 C^0_{org} + \ldots\ldots + R_n C^0_{org}\right)$$

$$-(L + R_1 + R_2 + \cdots\cdots + R_n)\ C^n_{org_{av}} = H_{ag} C^0_{aq_i}$$

or $$\left(L + \sum_{n=1}^{n} R_n\right)\left(C^0_{org} - C^n_{org_{av}}\right) = H_{ag} C^0_{aq_i} \qquad \text{...(18)}$$

Rate Equation

The mass transfer rate can be written as

$$H_{aq} dC^n_{aq_i} = \Re^n_{i_{av}} \left(C^n_{org_i} - \frac{C^n_{aq_i}}{\Re_D}\right)_{av} A_{drop} dv_n \qquad \text{...(19)}$$

Also, $$Lz_f \left(-dC^n_{org_i}\right) A_{drop} = \Re^n_{i_{av}} \left(C^n_{org_i} - \frac{C^n_{aq_i}}{\Re_D}\right)_{av} A_{drop} dv_n \qquad \text{...(20)}$$

and $$Lz_f \left(-dC^n_{org_i}\right) A_{drop} = \Re^n_{Torg_{av}} \left(C^n_{org_{av}} - C^n_{org_i}\right)_{av} A_{drop} dv_n \qquad \text{...(21)}$$

Note that the concentration gradients in equations 19, 20 and 21 are arithmetic average of the gradients between nth and $(n - 1)$th stage.

Mass Balance of solute for the 1st volume increment is given by

$$L\left(C^0_{org} - C^1_{org_{av}}\right) = H_{aq}\left(C^0_{aq_i} - C^1_{aq_i}\right) \qquad \text{...(22)}$$

Mathematical Criterion when Instability stops at the end of 1st stage[3]

The criterion is as follows:

$$\frac{M}{H} = 0.99 = \frac{RT\left(1 - \dfrac{C^1_{aq_i}}{C^1_{org_i} \Re_D}\right)}{\sigma^1_i S_o} \qquad \text{...(23)}$$

where,

$$\sigma^1_i = \sigma_{eq} - A + B \log \frac{C^1_{aq_i}}{C^1_{org_i}} \qquad \text{...(24)}$$

where σ_{eq}, A and B are constants for a ternary system[9].

Interfacial Mass Transfer Coefficient Correlation[5]

$$\text{Log } \Re^n_i = a - bC^0 - F'\sigma^n_i \qquad \text{...(25)}$$

where a, b and F' are constants for the specific ternary system.

Bulk Organic Phase Mass Transfer Coefficient Correlation[8]

The effect of density convection on mass transfer rate from bulk to interface has been taken into account by means of the following equation:

$$\frac{\Re_{T_{org}} h_{av}}{D_M} = 0.7(Sc)^{-1.04}(Ra)^{0.6} \qquad \text{...(26)}$$

where $Sc = \dfrac{\mu_{org}}{\rho_{org} D_M}$ and $Ra = \dfrac{\Delta\rho_{org} h_{av}^3 g}{\mu_{org} D_M}$

where $h_n = \dfrac{dv_n}{A_n}$ and $h_{av} = \dfrac{h_n}{2}$ since density eddies will rise from different heights within dv_n.

Mathematical criterion for start of instability and Mass Balance for injection of fresh feed at the end of 1st stage:

$$\frac{M}{H} = 1.01 = \frac{RT\left(1 - \dfrac{C^1_{aq_i}}{C^{1*}_{org_i} \Re_D}\right)}{\sigma_i^{1*} S_o} \qquad \text{...(27)}$$

$$\sigma_i^{1*} = \sigma_{eq} - A + B\log\frac{C^1_{aq_i}}{C^{1*}_{org_i}} \qquad \text{...(28)}$$

$$R_1 C^0_{org} = (R_1 + L) z_f A_{drop}\left(C^{1*}_{org_i} - C^1_{org_i}\right) \qquad \text{...(29)}$$

$$\text{Also, } C^{1*}_{org_{av}} = \left(LC^1_{org_{av}} + R_1 C^0_{org}\right)/(R_1 + L) \qquad \text{...(30)}$$

In the formulation of the design equation (29) it is assumed that fresh feed is introduced in the vicinity of the drops. The construction of the internal at this point in the column is done in such a manner that increase in concentration of solute in the interfacial region (z_f) is due to total amount of solute introduced through the fresh feed. The design of the internal is shown in Fig. 1.4.

Since, the drops will be distributed uniformly at any cross-section along the column one can make perforations in a plate and introduce organic phase surrounding each hole. The organic phase enters through a header and then gets injected at individual drop sites (perforations) (Figure 1.4). The perforations in the plate ought to be slightly bigger than the drop sizes so that the organic phase is injected in the film surrounding the drops thus increasing the concentration C_{org_i} such that σ_i decreases and hence instability is restored and mass transfer rate increases.

Mass Balance for the next stage:

Doing the mass balance for the next stage, one can write

$$(R_1 + L)\left(C^{1*}_{org_{av}} - C^2_{org_{av}}\right) = H_{aq}\left(C^1_{aq_i} - C^2_{aq_i}\right) \qquad \text{...(31)}$$

Rate Equation for the next stage:

$$H_{aq}\left(C^{1}_{aq_i} - C^{2}_{aq_i}\right) = \Re^{2}_{iav}\left(C^{1*}_{org_i} - C^{2}_{aq_i}/\Re_D\right)_{av} A_{drop} dv_2$$

$$= \Re^{2}_{T_{org_{av}}}\left(C^{1*}_{org_{av}} - C^{1*}_{org_i}\right)_{av} A_{drop} dv_2 \qquad ...(32)$$

Note that the concentration gradient is the arithmetic average of the gradient at the beginning of 2nd stage and that at the end of 2nd stage.

Mathematical criterion when instability stops in next stage:

As the concentration decreases the instability stops and one can write the criterion as,

$$\frac{M}{H} = 0.99 = RT\frac{\left(1 - \dfrac{C^{2}_{aq_i}}{C^{2}_{org_i}\Re_D}\right)}{\sigma^2_i S_o} \qquad ...(33)$$

where $\sigma^2_i = \sigma_{eq} - A + B\log\dfrac{C^{2}_{aq_i}}{C^{2}_{org_i}}$ ...(34)

Height of the *n*th Volume Increment:

The above equations (29 – 34) could be used along with equation (26) to determine the volume increment of the column for the section between 1 to 2, the 2nd volume increment after 1st side feed.

Thus, a general expression for height of *n*th stage can be written as,

$$h_n = dV_n/A_n \qquad ...(35)$$

where A_n is the cross-sectional area of the column between *n*th side feed and its next side feed and h_n is the height of this section. Height of 1st section can be obtained based on dv for 1st volume increment using eqns. (19) to (26) but with A_n for conventional column.

Continue with 2nd and 3rd side feed and so on up to the top of the column. The cross-sectional area A_n will vary in each zone from bottom to top. This is mainly to take care of increased flow of organic phase.

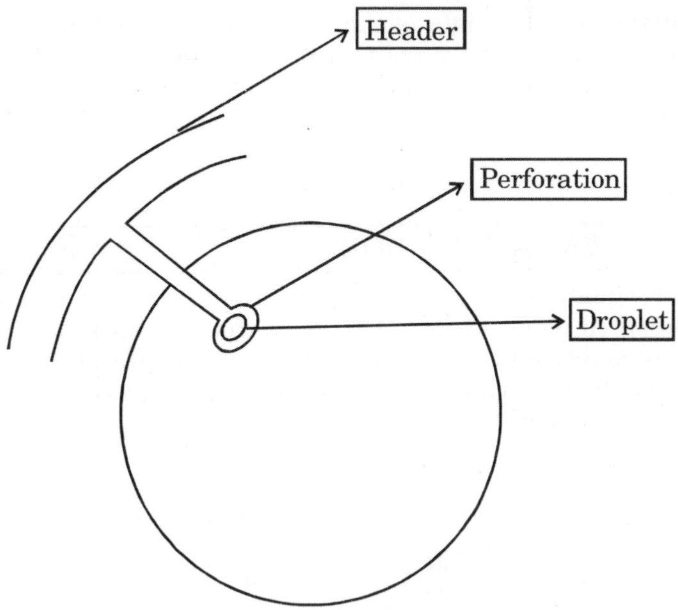

Figure 1.4: Feed inlet arrangement at any height above

Design Equation for Elemental Volume (dv)

The small volume element of any stage is shown in Fig. 1.5

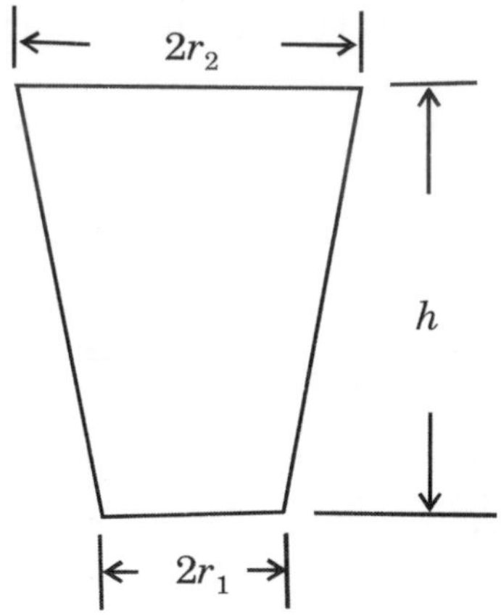

Figure 1.5: Tapered column

The volume of this type of tapered cylinder is given below:

$$dv_n = \frac{1}{3}\ \pi r_2^2 h_n + \frac{1}{3}\pi(r_2 + r_1)\ r_1 h_n \qquad ...(36)$$

where, r_1 is the radius of the base of the nth element, r_2 is that of the top section of that stage and h_n is the height of that stage.

During the calculation of the height of the nth stage the eqn. (36) is used instead of eqn. (35). Thus after evaluation of all the stage heights, one gets total column height (H_t) using summation.

$$H_t = \sum_{n=1}^{n} h_n \qquad ...(37)$$

1.4 SOLUTION OF DESIGN EQUATIONS

Solution of the design equations for multiple feed spray column has been done through a computer program prepared using TURBO-C^{++}. Note that in order to calculate height of the column one has to choose different values of M/H for the appearence and disappearence of energy instability. The following values of M/H have been choosen for these two states as follows and indicated in the text:

$$\left(\frac{M}{H}\right)_{starting} = 1.01$$

$$\left(\frac{M}{H}\right)_{stopping} = 0.99$$

Using the computer programme it has been found that for each volume of total assumed height the plot of the V and θ against M/H shows the same nature as follows:

In the plot of Fig. 1.6 and Fig. 1.7 optimized values of M/H is the point for which the two curves cross each other. From the plots it can be concluded that the optimized values of the M/H ratio for any value of the total height becomes nearer to 0.99 for lower range and 1.01 for higher range.

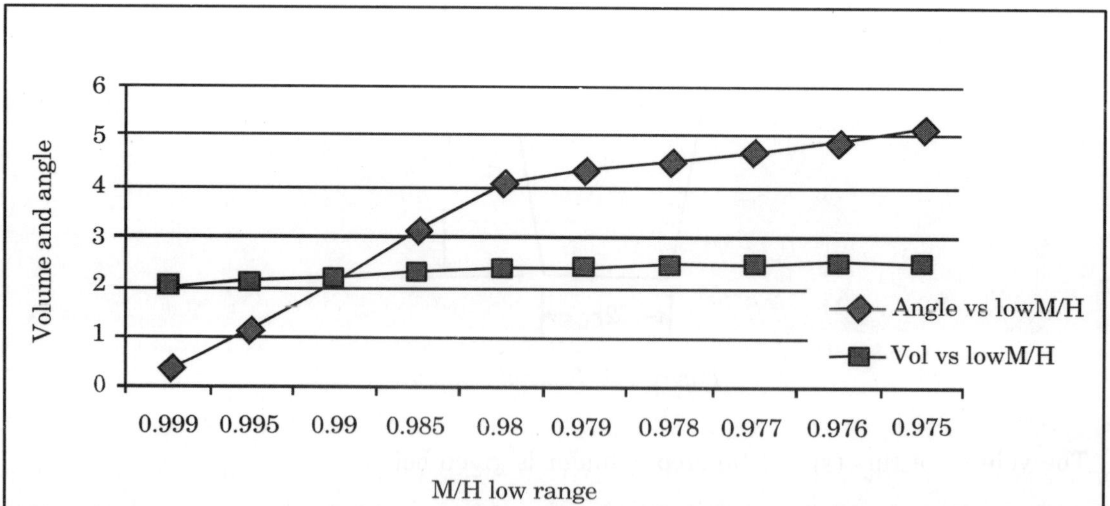

Figure 1.6: V and θ vs. M/H (Lower range, Instability stops) H = 0.707 m

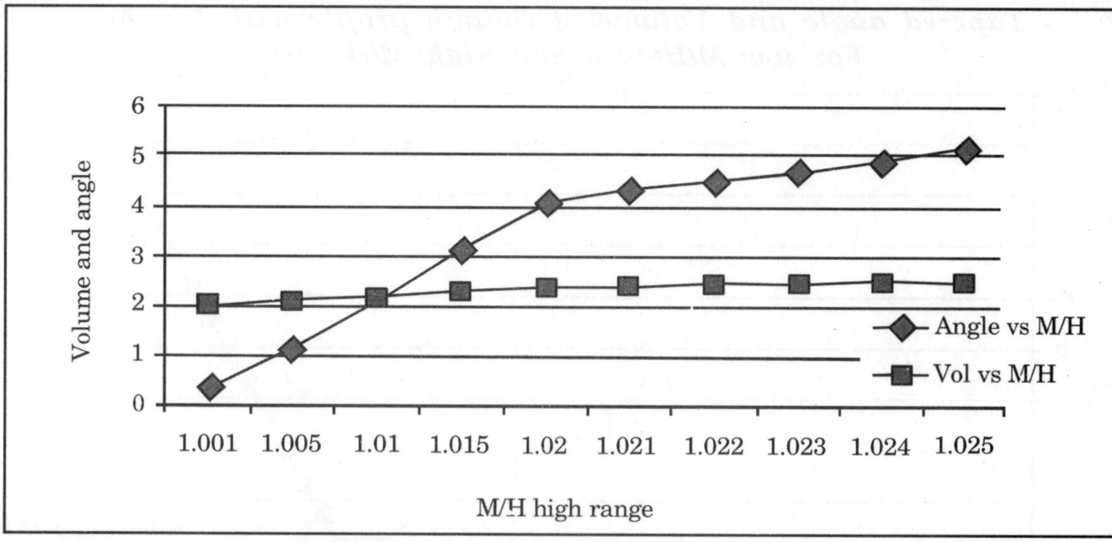

Figure 1.7: V and θ vs. M/H (Higher range, Instability starts) H = 0.707 m

1.5 VOLUME, HEIGHT AND TAPERING OF COLUMN

First of all it is assumed that the flux of organic phase is the same throughout the column. Therefore, for 1st stage (bottom)

$$\left(\frac{L+R_1}{\pi r_2^2}\right)=\left(\frac{L}{\pi r_1^2}\right)$$

where r_1 is the radius of the bottom most section of the column and L is the organic phase flow rate entering the bottom of the column. 'r_1' has been evaluated for a particular 'L' using flooding considerations for a cylindrical column of radius r_1 which is 0.6 meter (refer equation 38 & 39). Same flux is assumed throughout the column such that flooding correlation is valid. One may choose r_2 and therefore R_1 is fixed. Knowing R_1, dv_n is calculated from equations (29) to (34). Next, h_n is known from equation (36). Like this one can continue to solve for total height and get corresponding total volume of the column. By choosing different r_2 and corresponding R_1 one can generate a plot of total volume *vs* total height and the angle w.r.t. height and therefore from Figure 1.8 one can get the height and actual volume of the column from the intersection of the two plots (refer the example given below).

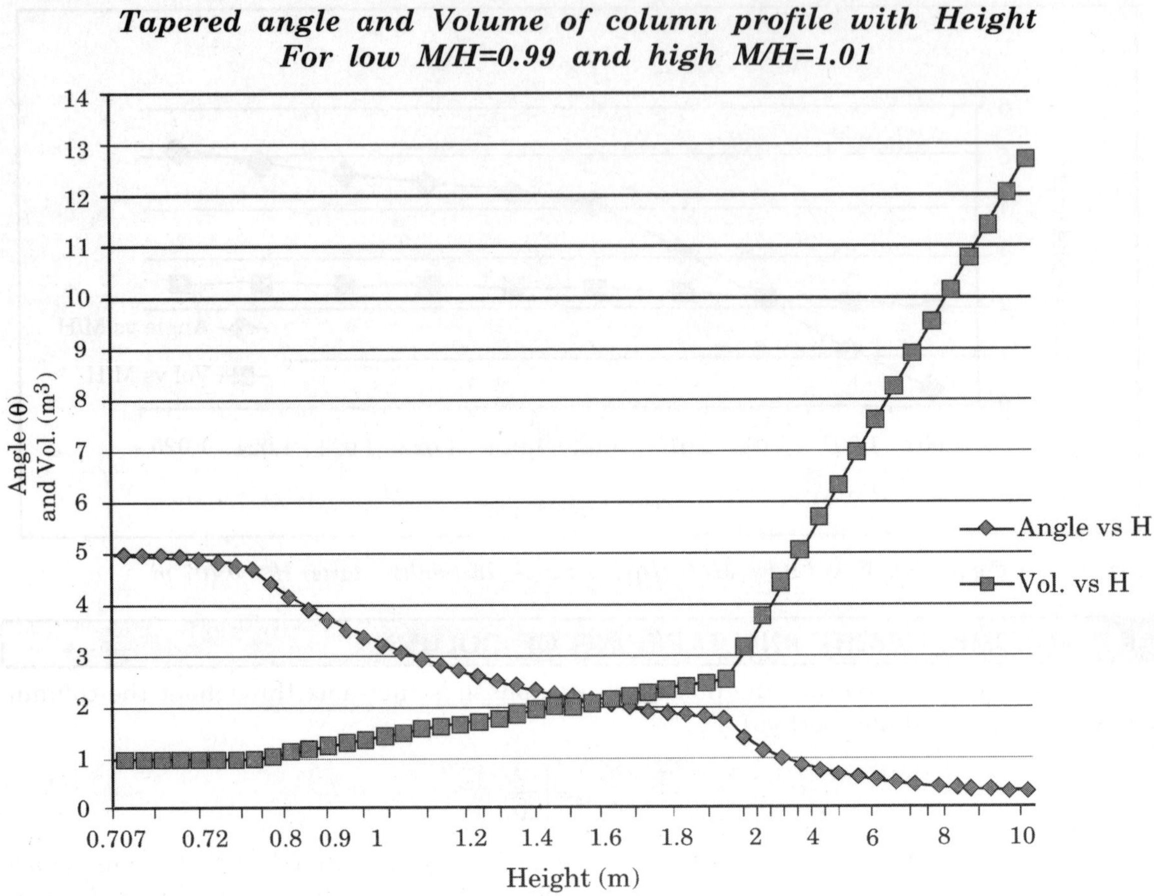

Figure 1.8: Tapered angle and volume of column profile with height

Conventional Design (11, 12, 13) (refer the example below):

1. Diameter of the column

Diameter has been estimated based on the following flooding velocity correlation:

$$V_{cf} = \frac{10{,}000\Delta\rho^{0.28}}{\left[0.453\mu_c^{0.075}\rho_c^{0.5} + d_p^{0.56}\rho_d^{0.5}\left(\frac{Q_d}{Q_c}\right)^{0.5}\right]^2} \quad \text{...(38)}$$

where Q_d and Q_c are dispersed and continuous phase flow rates under flooding condition and are equal. Note that equation (38) is based on FPS units.

μ_c = 4.33 lb/(ft)(hr), $\Delta\rho$ = 4.8 lb/ft^3, ρ_c = 55.92 lb/ft^3

d_p = 0.00328 ft, ρ_d = 60.72 lb/ft^3

$$d_t = \sqrt{\frac{4 \times L \times 3.28 \times 3600}{60 \times \pi \times 0.38 \times V_{cf}}} = 1.21 \text{ meter for 38\% of flooding velocity} \quad \text{...(39)}$$

2. Height of the column

Height = HTU × NTU

where $\text{HTU} = \dfrac{V_c}{K_{ca}}$ ...(40)

$$K_{ca} = \frac{\phi_d(1-\phi_d)\left(\dfrac{g^3\Delta\rho^3}{\sigma_i\rho_c^2}\right)^{0.25}}{(N_{sc,c})^{0.5} + \dfrac{1}{m}(N_{sc,d})^{0.5}} = 0.036 \text{ sec}^{-1} \quad ...(41)$$

where $\phi_d = 0.1$, $N_{sc,c} = 850$

$\Delta\rho = 80$ kg/m^3, $N_{sc,d} = 1055$

$\rho_c = 932$ kg/m^3

$m = 33$, $\sigma_I = 0.025$ N/m

$V_c = 0.03$ m/sec ∴ HTU = 0.83 m

$$\text{NTU} = \frac{C_{c_1} - C_{c_2}}{\Delta C_{C_{lm}}} = 11.52 \text{ where } C_{c_1} = 1.3 \text{ kgmole/m}^3, C_{c_2} = 0.013 \text{ kgmole/m}^3$$

$C_{d_1} = 30$ kgmole/m^3, $C_{d_2} = 0.0$ kgmole/m^3

$\Delta C_{c_1} = C_{c_1} - C_{d_1}/\text{m} = 0.391$ kgmole/m^3

$\Delta C_{c_2} = C_{c_2} - C_{d_2}/\text{m} = 0.0133$ kgmole/m^3

Height = 9.6 meter

1.6 RESULTS AND DISCUSSION

Based on the following example the results are shown in Figures 1.6 to 1.14

Acetic acid + toluene → H_2O

$C^0_{org} = 1300 \dfrac{\text{moles}}{\text{m}^3}$ $\Re_D = 33$

$C^0 = 13$ kgmole/m^3

$L = 2$ m^3/min $\sigma_{eq} = 24.6$, $A = 3.1$

$B = 2.75$

$C^0_{aq_i} = 30{,}000 \dfrac{\text{moles}}{\text{m}^3}$ $a = 10.2$, $b = 1.9$

$F' = 0.4$

T = 298° K

$z_f = \dfrac{1}{1000}$ meter $S_o = 0.0875 \times 10^{10}$ cm^2/mole

Normally, for carboxylic acids S_o is 0.125×10^{10} cm^2/mole.(3) But, in presence of the solvent toluene, it exists at the interface in associated form and hence S_o is different. 99% extraction of solute from organic to aqueous phase is assumed.

Change in concentration of solute in organic and aqueous phase with respect to height of the column has been studied. The behavior is logical i.e. decrease in concentration with respect to height (Figs. 1.9 & 1.10)

From Fig. 1.11 it is noticed that R decreases from bottom to top. This is because less quantity of organic phase is required to achieve M/H = 1.01 as one goes from bottom to top since $C^*_{org_i}$ decreases up the column.

From Fig. 1.12 it is clear that the dynamic interfacial tension gradually decreased from bottom to top. This is obvious from equation (24).

The degree of mixing of fresh feed at different height with the organic phase at the drop surface has been studied. A factor 'f' has been introduced in equation (29) as follows: $fR\ C^0_{org} = (fR + L)z_f A_{drop}\left(C^*_{org_i} - C_{org_i}\right)$. $f = 1$ means 100% mixing and $f < 1$ means lesser degree of mixing. Figure 1.13 shows how R increases as f decreases at different locations up the column.

Height of the spray column is 1.72 meters for multiple feed with instability assumption compared to 9.6 meters for conventional design procedure assuming no instability.

Figure 1.14 shows how k_i changes with height of the column. Note that k_i increases with height because σ_i decreases with height (Figure 1.12).

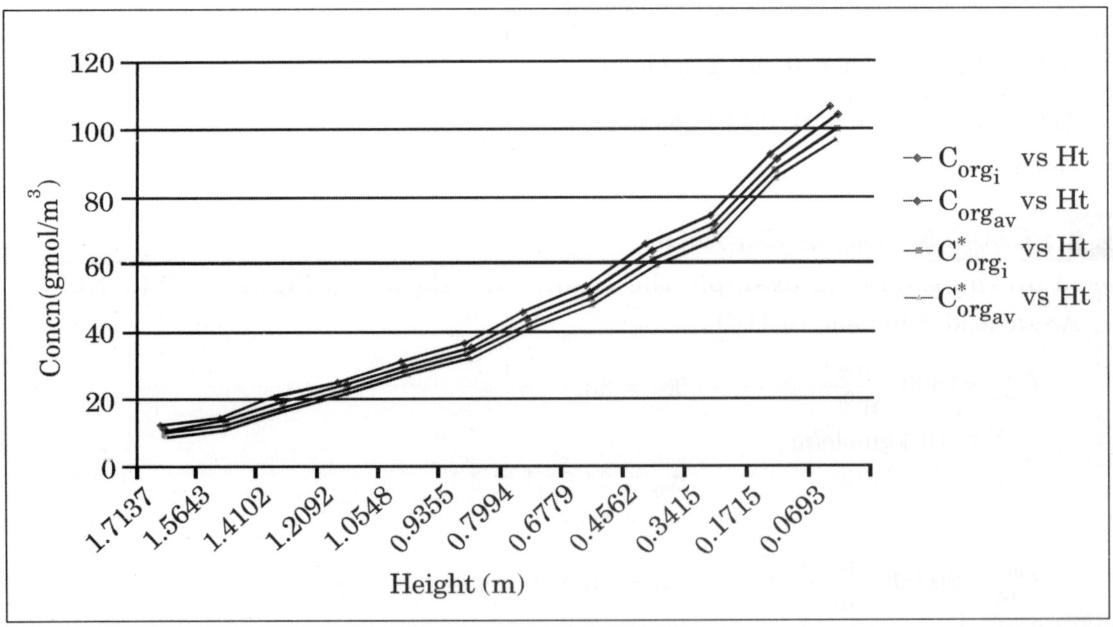

Figure 1.9: Interfacial and average concentration of solute in organic phase vs height

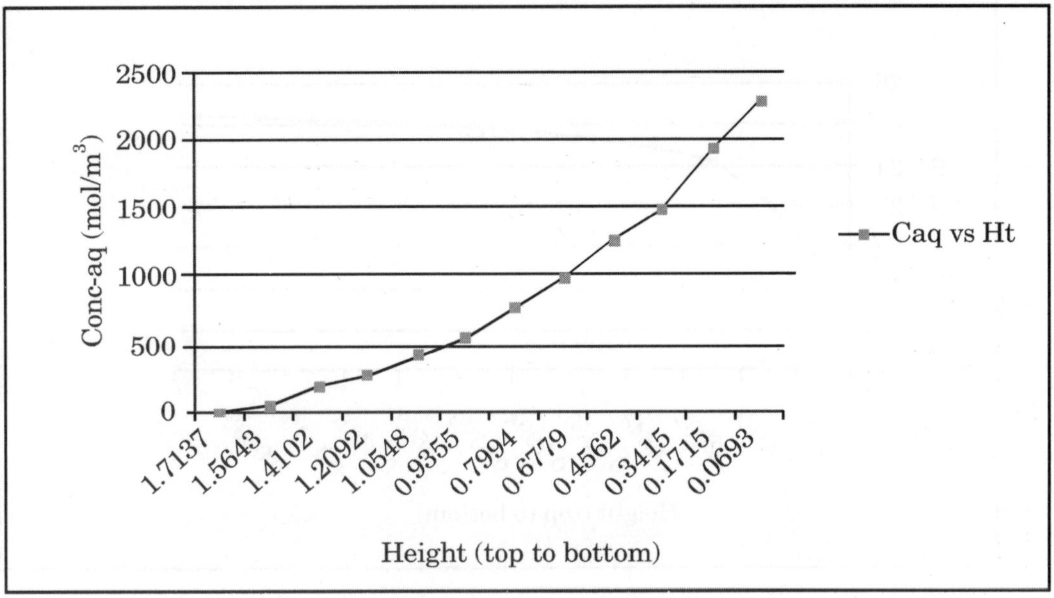

Figure 1.10: Concentration of solute in aqueous phase vs height

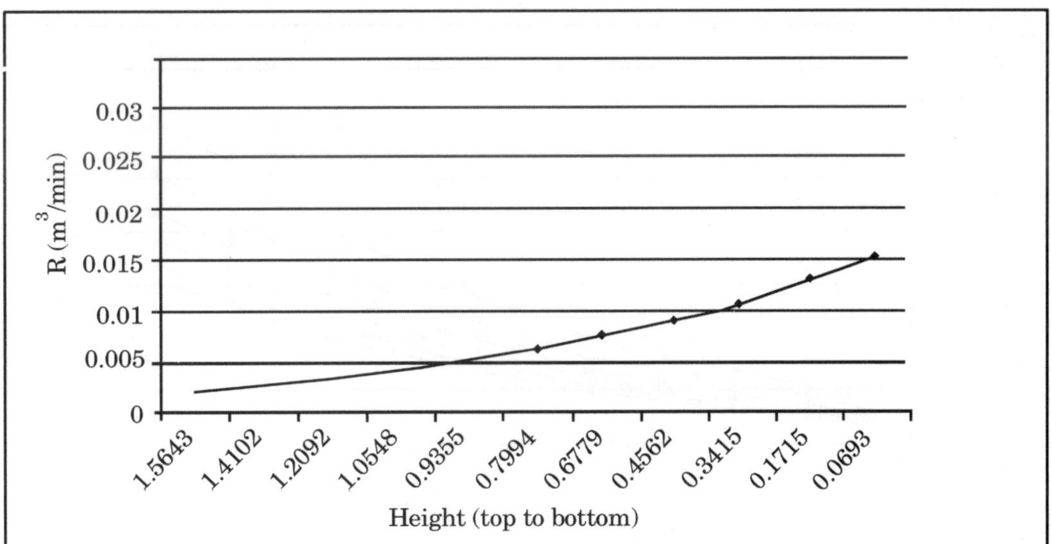

Figure 1.11: R vs. height for ëfí = 1

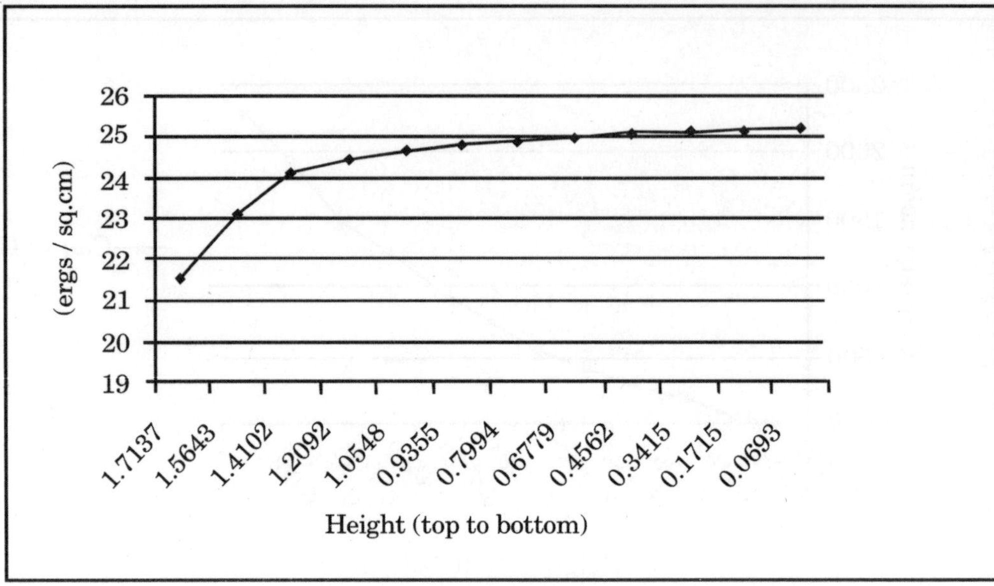

Figure 1.12: Dynamic interfacial tension vs. height

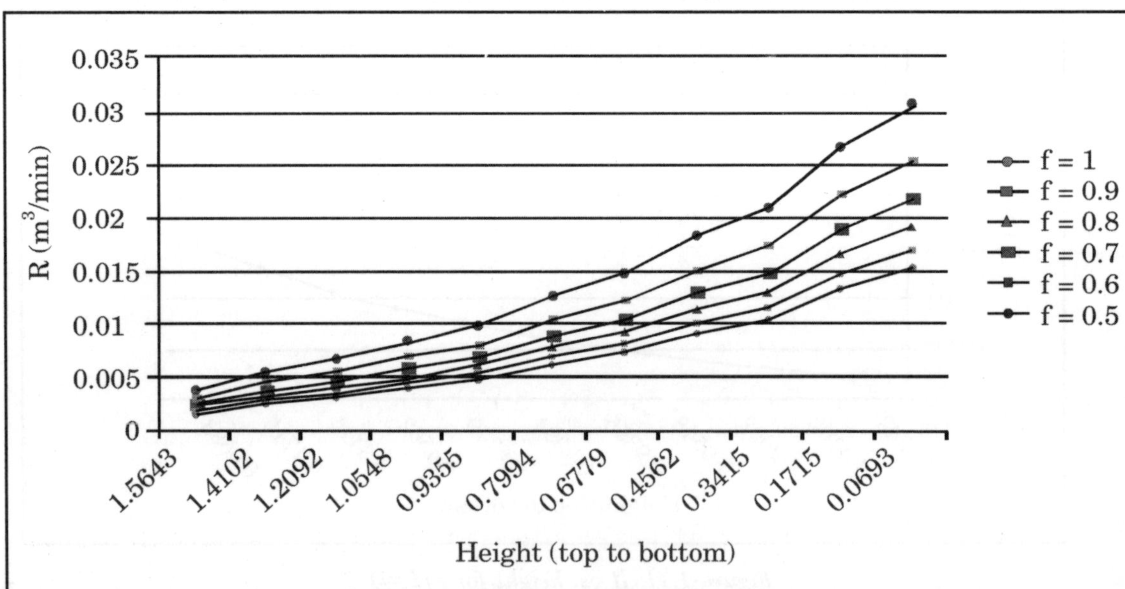

Figure 1.13: R vs. height for different ëfí

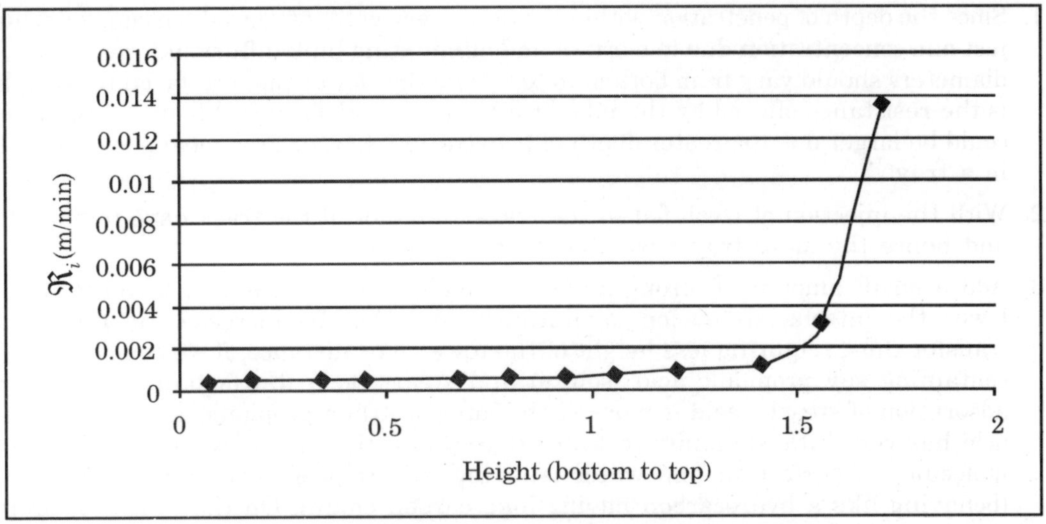

Figure 1.14: Interfacial mass transfer coefficient vs. height

Conclusion

It is obvious from the result that height of the column required for multiple feed is much less than that of the column with assumption of no instability. Furthermore, it is to be noted that with multiple feed the column height is less than single feed because in this case there will be alternate zones of instability and no instability as discussed earlier.

Power requirement for pump is maximum for the conventional column (with assumption of no instability) as because the height is much higher in this case. For single feed (with assumption of instability) power requirement is more than multiple feed because the column height is more. Only extra cost with multiple feed is (1) the cost of valves to control the desired flow at different height and necessary pipelines and (2) cost of internal design. However, material requirement is much less for the shell of the column because of tapered structure and less height.

1.7 SUGGESTION FOR FUTURE WORK

In spray columns there exists some disadvantage due to axial mixing which reduces the efficiency of such columns. A better design where axial mixing can be minimized would be an inverted normal sieve tray tower[16] where each tray would act as fresh feed injection tray and in the hold up above each tray contact between the heavy dispersed phase (the aqueous phase) and light continuous phase (the organic phase) would occur for mass transfer from organic to aqueous phase of desired amount. The trays may not be spaced equidistant from one another because of different volume required for liquid-liquid contact in different stage because of different interfacial tension (and hence $\Re_i$) and different concentration level, as one goes from bottom to top of the tower. For instance, at the top interfacial tension is minimum meaning high value of $\Re_i$ and concentration levels are lower thus, requiring less time of contact, meaning less hold up for desired mass transfer as one goes from bottom to top.

In addition to the extractor design suggested above which is an inverted normal sieve tray tower the following design considerations are recommended:

1. Since the depth of penetration within the drops vary with interfacial tension, for achieving just one concentration due to ejection and eliminating bulk diffusional resistance, drop diameters should vary from bottom to top. Note that lower the interfacial tension, lesser is the resistance offered by the interface to ejection activity and hence, drop diameter could be larger due to greater depth of penetration. Larger drop means less no. of holes in a tray.
2. With the injection of fresh fed in the trays diameter of the trays have to be different and hence the sieve tray tower has to be tapered one.
3. Add a small amount of strong surfactant e.g. stearic acid to any ternary system to lower the interfacial tension significantly and thereby increase the rate of mass transfer thus, requiring less height of the tower. For instance, if we consider a system containing say propionic acid (solute) and stearic acid (Surfactant) then degree of adsorption of stearic acid is more at the interface than propionic acid because stearic acid has very little solubility in water than propionic acid i.e. weak bonding between molecules of surfactant and water meaning that it prefers to stay at the interface (behaving like a hydrocarbon having long carbon chain). On the contrary, propionic acid has a high solubility in water meaning strong bonding between water and solute molecules and hence less adsorption at the interface.

   Surfactant molecules have a long hydrocarbon portion and a carboxyl group. They have greater affinity to stay adsorbed at the interface because the long hydrocarbon chain being organic in nature can stay at the interface in the organic side and the carboxyl group in the aqueous side. Such a situation is preferred because this way they can stay at minimum free energy level in comparison to staying soluble either in the organic or aqueous phase. Note that any system which has a high free energy level will always tend to come to equilibrium with minimum free energy, because that is a stable state according to thermodynamic principle.

   Reduction of interfacial tension is more in presence of surfactant because to begin with thay being a much larger molecule than surface active solute, the absolute energy level is higher and when they get adsorbed at the interface with minimum energy level, change in chemical potential or free energy is larger i.e. $\Delta\sigma$ is higher where $\Delta\sigma = \sigma_{eq} - \sigma_i$ (dynamic).

   With surface active solute reduction in interfacial tension is not that large because their absolute energy level is not that high, being a smaller molecule having a small alkyl group attached to the carboxyl group, and hence their affinity to remain adsorbed at the interface is not that great compared to surfactant.

   Thus $\Delta\sigma_{\text{surfactant}} > \Delta\sigma_{\text{surface active solute}}$

   It may be noted that a strong surfactant although reduces the interfacial tension, makes the interface stronger due to the strongly adsorbed monolayer. Thus, one must choose a particular critical concentration of strong surfactant such that physical resistance to ejection phenomenon is less compared to the degree of reduction of interfacial tension leading to higher rate of mass transfer. Physical resistance is due to viscous monolayer at the interface. Note that stearic acid is highly viscous. Highly viscous layer will offer physical resistance to penetration from one phase to another. Other strong surfactants are namely Lauric acid, Oleic acid etc. Experimentally, it has been found that critical concentration of the above mentioned surfactants vary in the range of 10^{-1} to 10^{-4} molar for occurrence of ejection phenomena possibly with high transfer rate(18) for various solute / solvent systems of varying initial concentration.

Additional mass transfer experiments need to be carried out to verify high transfer rate in a more convincing way.

Surfactant should be added both in the fresh feed at the bottom as well as along with the fresh feed in each tray. This is because at the end of 1st stage instability stops due to transfer of solute from organic to aqueous phase whereby interfacial tension rises and hence M/H decreases (<1). So in successive stages both fresh feed and surfactant need to be added to reduce the interfacial tension thereby restarting the instability (M/H becomes >1). Gradually less quantity of fresh feed and lower concentration of surfactant need to be added up the stages because required concentration of solute and surfactant decrease up the column in the vicinity of the drops (aqueous phase) for instability to reappear. This was evident from tower simulation results. If one does use only fresh feed containing solute and no surfactant in each tray then one would require large amount of fresh feed to reduce σ_i and yet mass transfer rate won't be high due to high activation energy. Thus, appropriate amount of strong surfactant needs to be added in each tray to reduce activation energy thus achieve a high transfer rate of solute throughout the tower. Activation energy is a function of dynamic interfacial tension and surfactant concentration.

1.8 FLUIDIZED BED EXTRACTOR FOR LEACHING

In fluidized bed extractor with stirrers at different levels the design of the stirrer is important, because the objective may be to tear apart the viscous slurry into large number of small fragments so that there occurs a large exposed surface for the solvent to extract maximum. If one uses a simple impeller or paddle or turbine then shearing occurs only at the tip of the impeller or paddle and along the length of it. A better design would be to make small holes within the impeller so that due to shearing action of the blades by the holes a large size fragment of viscous material, would be broken into small size fragments. Since the slurry is distributed throughout the fluidized bed, larger the diameter of the impeller better is the coverage. Also rpm of the stirrer should be high so that the shearing force will be of large magnitude. The stirrer consists of several numbers of rectangular bars attached to the shaft. Each bar has several small holes in it. The length of each bar should be about 90% of the radius of the extractor for maximum coverage. The thickness of the bar will depend on the viscosity of the slurry. If the viscosity is high thickness should be appreciable so that the bar does not bend due to resistance by the slurry. The holes should be of proper diameter so that maximum quantity of slurry is broken down to smaller sizes. However the clearance between the holes should be appreciable so that the small fragments are separated by sufficient void space containing solvent for proper extraction.

An example is extraction of $CaSO_4$ from coal particles by aqueous glycerol solution which is viscous. $CaSO_4$ is a poison to the catalyst, inherent in the mineral matter of coal which are responsible for coal liquefaction using hydrogen[10].

1.9 EXPERIMENTAL VERIFICATION OF THE DESIGN CONCEPTS DISCUSSED IN THIS CHAPTER

Appropriate experiments have been carried out to verify the design concepts discussed in this chapter and it has been found that the results are satisfactory, and corroborate the theory behind the design concepts.

NOMENCLATURE

A, B	–	constant (function of specific ternary system) in ergs/cm^2
A'	–	constant (function of C^0_{org}) in decimeter/min
A_{drop}	–	surface area of drops per unit volume of the column in m^2/m^3
A_n	–	cross-sectional area of nth section of column in m^2
$C1_i$	–	interfacial concentration of solute in phase 1 from which solute is transferred to phase 2 in moles/m^3
$C2_i$	–	interfacial concentration of solute in phase 2 to which solute is transferred from phase 1 in moles/m^3
$C^n_{aq_i}$	–	interfacial concentration of solute in aqueous phase in the nth stage in moles/m^3
$C_{aq_{av}}$	–	average concentration of solute in aqueous phase in the region of column where there is no instability in moles/m^3
C_{aq}	–	concentration of solute in aqueous phase at any radius inside the drop in moles/m^3
C_{c_1}	–	initial concentration of solute in organic phase in kgmole/m^3
C_{c_2}	–	outlet concentration of solute in organic phase in kgmole/m^3
C_{d_1}	–	outlet concentration of solute in extract in kgmole/m^3
C_{d_2}	–	initial concentration of solute in extract in kgmole/m^3
$\Delta C_{c_{lm}}$	–	log mean driving force in kgmole/m^3
$C^n_{org_i}$	–	interfacial concentration of solute in organic phase in the nth stage in moles/m^3
$C^n_{org_{av}}$	–	average concentration of solute in organic phase in the nth stage in moles/m^3
C^0_{org}	–	concentration of solute in the organic phase in the feed to the column in moles/m^3
C^0	–	initial concentration of solute in the feed to the column in organic phase in kgmole/m^3
$C^{n^*}_{org_i}$	–	concentration of solute just outside the drop surface after restoration of instability in case of multiple feed spray column in moles/m^3
d_p	–	diameter of the spherical aqueous phase droplets in ft
d_t	–	spray column diameter in meter
D_M	–	molecular diffusivity of solute in organic solvent in m^2/sec
D_W	–	molecular diffusivity of solute in aqueous phase in m^2/sec

dv_n – incremental volume of spray column in m^3 for nth stage

F' – a constant specific for a ternary system

g – acceleration due to gravity in m/sec^2

H – free energy of the interface in ergs/mole

H_t – total column height in meter

H_{aq} – aqueous phase flow rate in m^3/min

h_n – height of column corresponding to the incremental volume dv_n in meter

K_{CA} – overall mass transfer capacity coefficient based on the continuous phase in sec^{-1}

$\mathfrak{R}^n_{T_{org_{av}}}$ – arithmetic average of convective mass transfer coefficient of solute in the organic phase between nth and (n-1)th stage in m/min

$\mathfrak{R}_{T_{org}}$ – convective mass transfer coefficient of solute in the organic phase in m/sec

$K^n_{i_{av}}$ – arithmetic average of interfacial mass transfer coefficient in meter per minute between nth and (n – 1)th stage.

$\mathfrak{R}^n_i$ – interfacial mass transfer coefficient in decimeter per minute for nth stage

$\mathfrak{R}_D$ – distribution coefficient

K_D – interfacial mass transfer coefficient in absence of instability in decimeter/min

L – organic phase flow rate in m^3/min

M – specific energy of driving force of mass transfer in ergs/mole

m – distribution coefficient

n – no. of drops in dv

N_{Sc}, c – Schmidt number – continuous phase

N_{Sc}, d – Schmidt number – dispersed phase

r – any radius within drop in meter

r_p – radius of drop in meter

r_1, r_2 – tapered column radii in meter

R – universal gas constant

Ra – Rayleigh number for organic phase

R_n – fresh feed injection in nth stage in mole/m^3

S_0 – area occupied by a mole of solute at the interface in cm^2/mole

S_c – Schmidt number for organic phase

T – absolute temperature of extraction °K

V_c – continuous phase velocity in m/sec

V_{cf} – continuous phase flooding velocity in ft/hr

z_f ñ organic phase film thickness from which mass transfer occurs into aqueous phase due to interfacial instability in meter

Greek Letters

ρ_{org}	–	organic phase density in kg/m^3
ρ_c	–	density of continuous phase in lbs/ft^3, kg/m^3
ρ_d	–	density of dispersed phase in lbs/ft^3
μ_c	–	viscosity of continuous phase in lb/(ft)(hr)
μ_{org}	–	organic phase viscosity in kg/(m) (sec)
$\Delta\rho$	–	density difference between the two phases in lb/ft^3, kg/m^3
$\Delta\rho_{org}$	–	difference in density between the organic phase near the interface and that w.r.t. average solute concentration in kg/m^3
ϕ_d	–	dispersed phase hold-up
σ_i^n	–	dynamic interfacial tension in $ergs/cm^2$ for the *n*th stage when instability stops
$\sigma_i^{n^*}$	–	dynamic interfacial tension in $ergs/cm^2$ for *n*th stage when instability is restored.
σ_{eq}	–	equilibrium interfacial tension in $ergs/cm^2$

REFERENCES

1. Sawistowski, H., Interfacial Phenomena (chapter 9), *Recent Advances in Liquid-Liquid Extraction;* edited by Hanson,C., Pergamon Press, 1971.
2. Sternling, C.V., Acriven, L.E., AIChE Journal, 5, 514 (1959).
3. Abramzon, A.A., Kremnev, L.Ya., Frumin, G.T., and Ostrovskii, M.V., Zhurnal Prikladnoi Khimii, 40, 1319 (1967).
4. Sirkar, A., Journal of the Institution of Engineers (India), 61 pt Chapter 3, 1981, p. 62.
5. Abramzon, A.A., Ostrovskii, M.V. and Frumin, G.T., Zhurnal Prikladnoi Khimii, 41, 803 (1968).
6. Berg, J. C. and Morig, C.R., Chem. Engg. Sci. 24, 1969, p. 937.
7. Walas, S.M., Sirkar, A., Chem. Engg. Sci. 4, 1979, p. 741.
8. Sirkar, A., Manna, T.K., and Paruya, S., Paper presented at CHEMCON 2003 held at Bhubaneswar in December 2003.
9. Same as ref 3, 40, (1967), p. 1328.
10. Sirkar, A. Indian Patent 235876 (2009).
11. Thorton, Chem. Engg. Sci, 5, 1956, p. 201.
12. Minard and Johnson, Chem. Engg. Prog. 48, 1952, p. 62.
13. Cavers in Lo et al. Handbook of solvent Extraction, John & Sons (New Work), 1983 pp. 323 and 327.
14. Sirkar, A. Indian Patent 189418 (2004).
15. Sirkar, A. Indian Patent 194454 (2005).
16. Treybal, R.E. Mass Transfer operations, 3rd edition, Mc Graw-Hill 1980.
17. England, E.C. and Berg, J.C. AIChE Journal, vol. 17, No. 2, 1971, p. 313.
18. Abramzon, A.A., Ostrovskii, M.V., Frumin, G.T., Zhurnal Prikladnoi Khimii, vol 41, No. 9, 1968, pp. 1993.

Section II

CHAPTER 2

Reactor Design

Four types of reactor design have been discussed in this chapter as follows:

1. Preheater-Cum-Reactor for Coal Liquefaction based on Dry Hydrogenation[1]
2. Moving Bed Type Preheater for naphtha
3. Reactor for Liquid Phase Hydrogenation of Coal
4. Gas-Solid Catalytic Reactor Design using non-conventional shape catalyst with vibratory movement of the reactor.

For type 1 and 2 discussion is restricted to various preheater design concepts and for type 4 discussion is restricted to development of a mathematical model for such type of reactors. For type 3 discussion is mainly with reference to scale-up of such reactors.

2.1 DRY HYDROGENATION OF COAL

Description of Preheater Cum-Reactor Design

In consists of a rectangular vessel (Figure 2.1) with pressurized coal feed from top at one end and hot hydrogen enters from the same end, just below the coal entry point. The hydrogen entrains the coal particles and itself being a free jet, creates enough turbulence through the entertainment of surrounding fluid (Hydrogen) and coal thus, there occurs enough mixing between coal and hydrogen and the desired operating temperature is reached in 0.001 second. The jet hits the opposite end of the preheater and bounces back on the same side from where hydrogen entered. Only two paths have been considered for design purpose in order to fulfill the time constraint of 0.001 second. The mixture of coal and hydrogen next passes through a pipeline reactor where desired residence time is to be given for maximum yield of tar and liquid. Preheating time is restricted to 0.001 second in order that coal passes through the plastic region[2] (325 to 350°C) quickly. Otherwise, the plastic coal may stick to the wall and jeopardise the liquefaction process.

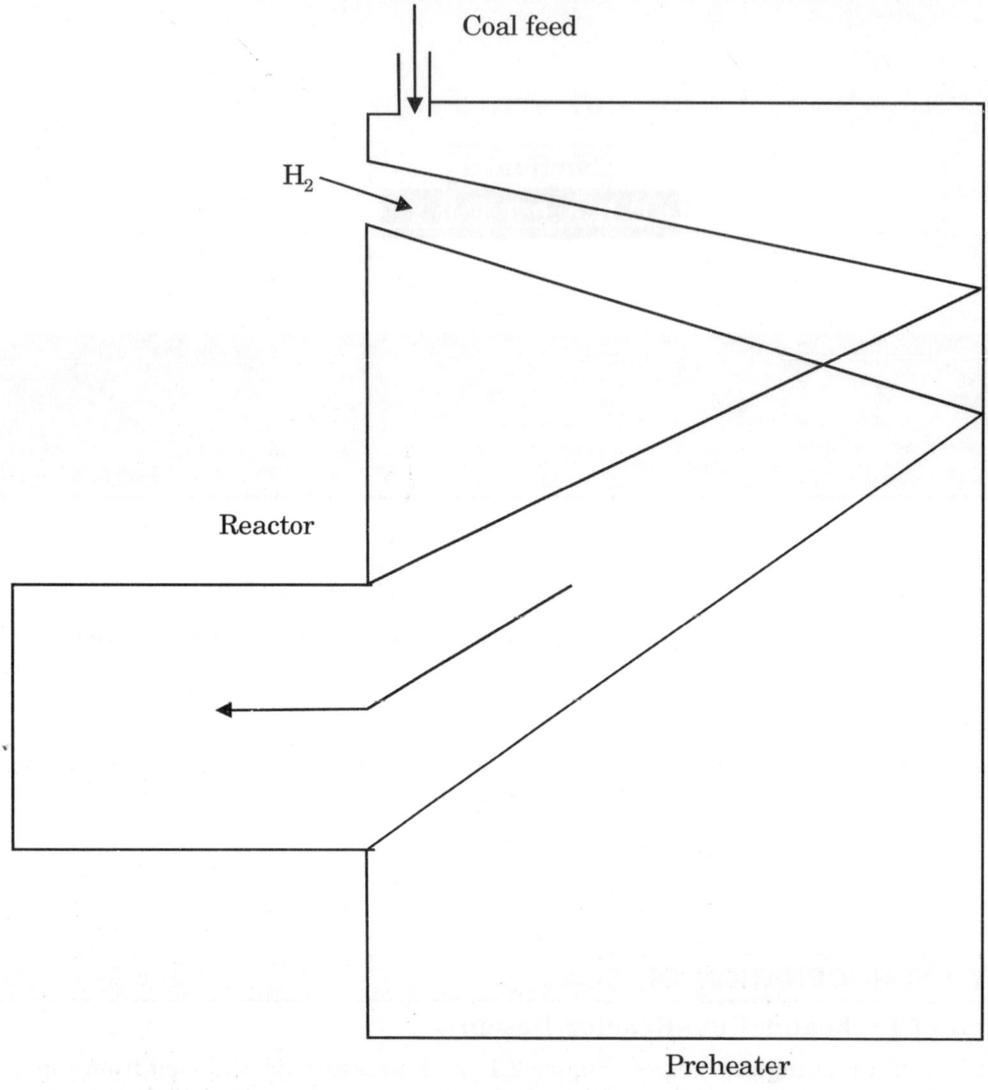

Figure 2.1: Preheater cum reactor

DESIGN BASIS

(i) Jet bounces back

Initially, jet is a mixture of hydrogen gas and coal particles. It is assumed that coal passes through the plastic region after coal and hydrogen mixture hits the opposite wall of the preheater. Since, the gas is elastic in nature its coefficient of elasticity 'e' will be close to unity. A value of 0.9 for 'e' has been assumed for calculation purpose. But, the coal particles' coefficient of elasticity is different from 0.9. However, because coal is being entrained by hydrogen, when hydrogen bounces back it will entrain bounced back coal particles along with it and the combined stream's angle of reflectance θ_2 is related to θ_1 by the following equation[(3)]:

$\cot \theta_2 = e \text{ Cot } \theta_1$

where 'e' has been taken as 0.9.

Also, $V_{H_{22}}$ (Average velocity of the bounced jet) is given by[3]

$$V_{H_{22}} = V_{H_{21}} \sqrt{\sin^2 \theta_1 + e^2 \cos^2 \theta_1}.$$

(ii) Heat transfer

In absence of any correlation on heat transfer coefficient for free jet to solid particles it is assumed that the hydrogen jet's temperature varies linearly along the jet path. It is necessary to know the jet temperature profile in order to calculate 'α', the jet spread angle which is a function of μ/ρ for the expanded jet region where flow is not fully developed[4]. This is given below. In the fully developed flow region 'α' is constant.

$\tan \alpha = 0.238 \ (\mu/\rho)^{0.135}$

At the later part of this chapter it has been shown that assuming fixed bed heat transfer coefficient correlation, heat supplied to coal particles is greater than the heat required to preheat coal, thus, satisfying our design calculations based on fluid dynamics aspects of preheater design.

(iii) Jet residence time is 0.001 second

The reasons have been elaborated earlier.

(iv) Only two jet paths

Only two jet paths have been considered. This is due to the fact that time of contact is 1 millisecond and the inlet velocity of hydrogen is kept below sonic region. If jet travels more than two paths then the contact time exceeds 1 millisecond.

(v) Feed rate

The feed rate is 1 ton of coal per day and the size is such that 70% of it passes through 200 mesh screen. This is the particle size used by previous workers[2].

(vi) Free jet turbulence

A free jet, upon leaving an outlet, will entrain the surrounding fluid, expand and decelerate. To a first approximation, total momentum is conserved as jet momentum is transferred to the entrained fluid. For practical purposes, a jet is considered free when its cross-sectional area is less than one-fifth of the total cross-sectional flow area of the region through which the jet is flowing. A turbulent jet in this discussion is considered to be a free jet with Reynolds number greater than 2000[5]. In the present case initial jet Reynold's No. has been found to be greater than 2000 within the range of parameters used for design calculation. The ranges of parameters are as follows:

1. Initial velocity of hydrogen – 125 to 300 m/sec
2. Initial temperature of hydrogen – 500 to 650°C

For the free jet there exists two region; first is the region where flow is not fully developed and the second zone is where flow is fully developed. For the first region the following relationship between average jet velocity w.r.t. jet centerline distance from entrance has been developed using the mass balance equation and data reported in a paper by Hill[6].

$$\overline{u}_x = \frac{0.1045}{D_0^3} \times \frac{(m_1 x + A_2)^{0.73}}{(m_1 x + A_1)^{0.27}} \times m_{H_2} \times \left\{0.0325 \frac{D_0^2}{x} + 0.0294 D_0 - 0.00093x\right\}$$

for $0 < \frac{x}{D_0} \leq 7$

where, $A_1 = T_{H_{2in}} + 940$

$A_2 = T_{H_{2in}} + 21.5$

$m_1 = \left(T_{H_{2out}} - T_{H_{2in}}\right) / (x_1 + x_2)$

$T_{H_{2out}} = 723°K$

$$m_{H_2} = \frac{0.01157 \times 1008 \times 425}{\frac{0.0179}{2}\left(T_{H_{2in}} + T_{H_{2out}}\right) \times 1000 \times \left(T_{H_{2in}} - T_{H_{2out}}\right)}$$

Here $C_{p_{coal}} = 1008 \frac{J}{(kg)(°K)}$

and $C_{pH_2 av} = \frac{0.0179}{2}\left(T_{H_{2in}} + T_{H_{2out}}\right) \times 1000 \frac{J}{(kg)(°K)}$

and $D_0 = \sqrt{\frac{m_{H_2} \times v_{H_2} \times 1000 \times 4}{V_{H_{2i}} \times \pi \times 2 \times 10^6}}$

where v_{H_2} is specific volume of H_2 in $\frac{cc}{gmole}$ and is equal to $170 + 0.625\left(T_{H_{2in}} - 273\right)$.

For the region where flow is fully developed the centerline velocity of decay is given by[(5)]

$V_c / V_o = 1.41\ Re^{0.135} (D_o/x)$ for $7 < x/D_o < 100$

where $Re = \frac{D_o V_o \rho_{H_2}}{\mu_{H_2}}$ is the initial jet Reynold's number

Also, radial velocity distribution is given by[(5)]

$\text{Log}\ (V_c/V_r) = 40\ (r/x)^2$ for $7 < x/D_o < 100$

(vii) Coal preheated to 450°C

It has been assumed that coal be preheated to 450°C within one millisecond. Previous experiments[(2)] suggest that for maximum yield of tar and liquid operating temperature should be 450 to 500°C. 450°C has been chosen because it was observed that with decrease in temperature yields of both tar and liquid increase[(2)].

(viii) Angle of incidence

The angle of incidence of the jet (θ_1) is assumed to be slightly greater than 'α'. Because the jet spreads, if one takes the angle to be α or less than α, then some portion of the fluid (those on the upper boundary) would bounce back vertically leading to mixed jet path in the preheater. This is not desirable as because one would like the jet to follow a particular path leading to controlled residence time. Thus, the incident jet angle θ_1 has been taken as 4° since 'α' varies between 2.5 to 3.0° within the range of temperature used in the calculation.

(ix) Hydrogen properties

Hydrogen inlet pressure is 2000 psi(2) and inlet temperature ranges from 500 to 650°C. This is equivalent to 50,000 to 2,00,000 °C / sec heating rate of coal. The hydrogen properties were evaluated using thermodynamic data available in Perry's Handbook(5) and Smith's book(7).

(x) Coal enters at 2000 psi and at an ambient temperature of 25°C.

Design equations for the multipass preheater

Since, over a portion of the forward path flow is not fully developed and for the rest flow is fully developed, different equations for average jet velocity are to be used as follows:

For x up to $7D_o$, the average incident jet velocity is given by $\overline{u_x}$ referred earlier in the text (refer item *vi* under design basis). For the portion of forward path beyond $x = 7D_o$ the average incident jet velocity is given by

$$V_{H_{21}} = \frac{V_o K D_o (1 - e^{-92.1 \tan^2 \alpha_1})}{92.1 x_1 \tan^2 \alpha_1} \qquad \text{...(1)}$$

This expression has been obtained using the equations for V_c/V_o and V_c/V_r given in item (*vi*) under design basis.

$$V_{H_{22}} = \frac{V_{H_{21}} K (D_{01} + D_{02})(1 - e^{-92.1 \tan^2 \alpha_1})}{92.1 x_2 \tan^2 \alpha_1} \sqrt{\sin^2 \theta_1 + e^2 \cos^2 \theta_1} \qquad \text{...(2)}$$

where $\theta_1 = 4°$ (defind earlier in the text)

$$\alpha_1 = 0.238 \left(\frac{\mu}{\rho}\right)^{0.135} \quad \text{at } x_1 \geq 7D_o$$

Total heating time = 0.001 second = $t_1 + t_2 + t_3$

where t_1 = contact time for the region of forward path where flow is not fully developed

$$t_1 = \frac{306.5 D_o^3 A_1^{0.27}}{m_{H_2} A_2^{0.73}} \left\{ 2.48 - 123.113 D_o \left(\frac{0.27 m_1}{A_1} - \frac{0.73 m_1}{A_2} \right) + \frac{0.197 m_1^2}{A_1 A_2} \left(4607.196 D_o^2 \right) \right\} \qquad \text{...(3)}$$

where A_1, A_2 and m_1 are defined as follows:

$$A_1 = T_{H_{2in}} + 940$$

$$A_2 = T_{H_{2in}} + 21.5$$

$$m_1 = \left(T_{H_{2out}} - T_{H_{2in}}\right)/(x_1 + x_2)$$

$$T_{H_{2out}} = 723^\circ K$$

t_2 = contact time for the region of forward path where flow is fully developed

$$= \frac{92.1\left[x_1^2 - y_1^2\right]\tan^2\alpha_1}{2V_o KD_o(1-e^{-92.1\tan^2\alpha_1})} \quad ...(4)$$

t_3 = contact time for the backward path of the bounced jet

$$= \frac{92.1x_2^2\tan^2\alpha_1}{2V_{H_{21}}K(D_{01}+D_{02})(1-e^{-92.1\tan^2\alpha_1})\sqrt{\sin^2\theta_1 + e^2\cos^2\theta_1}} \quad ...(5)$$

Volume of preheater = $s \times w \times h$

where w = Width of preheater

= $D_{03} + D_{04}$ + clearance

Assume clearance = 0.01 meter at both ends

D_{01}, D_{02}, D_3 and D_{04} are shown in Figure 2.2 and are given by the following expressions:

$$D_{01} = D_o/2 + \{1 + \tan\theta_1 \tan(\theta_1 - \alpha_1)\}\ s.\tan\alpha_1 \quad ...(6)$$

$$D_{02} = D_o/2 + \{1 + \tan\theta_1 \tan(\theta_1 + \alpha_1)\}\ s.\tan\alpha_1 \quad ...(7)$$

$$D_{011} = D_{01} + D_{02}$$

$$D_{03} = D_{01} + \{1 + \tan\theta_2 \tan(\theta_2 - \alpha_1)\}\ s.\tan\alpha_1 \quad ...(8)$$

$$D_{04} = D_{02} + \{1 + \tan\theta_2 \tan(\theta_2 + \alpha_1)\}\ s.\tan\alpha_1 \quad ...(9)$$

$$D_{022} = D_{03} + D_{04}$$

h = Height of preheater = C_1(Clearance above H_2 jet inlet) + $D_o/2$ + s $\tan\theta_1$ + s $\tan\theta_2$ + D_{o4} + C_2 (clearance below outgoing jet)

where clearance above H_2 jet inlet $= \frac{5\pi D_o^2}{8w}$ (refer item *vi* under design basis)

Assume clearance below outgoing jet = 0.01 meter. Also, $s = x_1\cos\theta_1 = x_2\cos\theta_2$

Printout of two computer programs developed to calculate volume of preheater for different initial hydrogen velocity and inlet hydrogen temperature are given below.

1. Main program, where total time of jet travel and x_1 and x_2 are calculated satisfying the error criteria (see at the end of listing).
2. For the case which satisfies the above criteria volume is calculated by the volume program.

(1) Main program

```
# include<stdio.h>
# include<math.h>
# include<conio.h>
```

```
main ( )
{

float T_H2_IN,A1,A2,T_H2_OUT,m1,Xf,X1,X2,D0,y1,M_H2,V0,K,Re;
float  time_total,time_total_calculated,T_calu,time_1,time_2,time_3;
float
INT_VAR1,INT_VAR2,INT_VAR3,INT_VAR4,INT_VAR5,INT_VAR6,INT_VAR7,
INT_VAR8,INT_VAR9,INT_VAR10;
float
INT_VAR11,INT_VAR12,INT_VAR13,INT_VAR14,INT_VAR15,INT_VAR16,INT_VAR17,
INT_VAR18,INT_VAR19,INT_VAR20,Diff;

float  tolerance,tolerance2;

float M_COAL,V_H2_I,CP_H2_AV,CP_COAL;
float a,b,x,Miu,sp_vol,Mass_H2,row_H2,Miu_by_row,V_H2r,ds;
float VH_21,D_01,D_02,s,ALFA_1,THEATA_1,w,h,v;
float  Miu_c,  sp_vol_c,row_H2_c,Miu_by_row_c,D_03,D_04,e1;
FILE *fp;
fp = fopen("preheater.txt","w");
// Calculation of Tem

T_H2_IN = 500+273;
A1 = T_H2_IN+940;
A2 = T_H2_IN+21.5;
e1 = 0.9;
THEATA_1 = 4;
T_H2_OUT = 723.0;

M_COAL = 0.01157;
CP_H2_AV = (0.0179*(T_H2_IN+ T_H2_OUT)*1000/2);
CP_COAL = 1008.0;
M_H2 = ((M_COAL*CP_COAL*425)/(CP_H2_AV*(T_H2_IN- T_H2_OUT)));

printf("Enter the value of V_H2_I(m/s)= ");
scanf("%f",&V_H2_I);
V0 = (M_H2*V_H2_I)/(M_H2+M_COAL);
```

```
//Miu_by_row at the inlet temp
Miu = (0.000000015*T_H2_IN+0.000005031);
sp_vol = (0.0006*T_H2_IN+0.0165);
Mass_H2 = 2;
row_H2 = Mass_H2/sp_vol;
Miu_by_row = Miu/row_H2;

// Dia of inlet jet
D0 = sqrt((M_H2*sp_vol*4)/(V_H2_I*3.1416*2));
fprintf(fp,"DO = %f",D0);
Re = (D0*V0/Miu_by_row);

K = 1.41*pow(Re,0.135);
time_total = 0.001;
y1 = 7*D0;


printf("\n t_t_c \t Xf \t X_12 \t Diff \t X1 \t X2 \t time_1 \t time_2 \t time_3");
fprintf(fp,"\n t_t_c \t Xf \t X_12 \t Diff \t X1 \t X2 \t time_1 \t time_2 \t time_3");

for(Xf = 10*D0;Xf<0.5;Xf=Xf+0.00001)
{

    m1=(T_H2_OUT-T_H2_IN)/(Xf);
    INT_VAR1 = (306.5*D0*D0*D0*(pow(A1,0.27))/(M_H2*(pow(A2,0.73))));
    INT_VAR2 = (2.48-123.113*D0*((0.27*m1/A1)-(0.73*m1/A2)));
    INT_VAR3 = 0.197*m1*m1*(4607.196*D0*D0)/(A1*A2);

    time_1 = INT_VAR1*(INT_VAR2+INT_VAR3);

    for(X1 = 8*D0;X1<0.25;X1=X1+0.00001)
        {
                // T = a+b.x
                a = T_H2_IN;
                b = (T_H2_OUT-T_H2_IN)/Xf;
                T_calu = a+b*7*D0;
```

```
Miu_c = (0.000000015*T_calu+0.000005031);
sp_vol_c = (0.0006*T_calu+0.0165);
row_H2_c = Mass_H2/sp_vol_c;
Miu_by_row_c = Miu_c/row_H2_c;
//code changed
ALFA_1 = 0.238*(pow(Miu_by_row_c,0.135))*(180/3.1416);
s = X1*(cos(THEATA_1*(3.1416/180)));

INT_VAR4 = 92.1*(tan(ALFA_1*(3.1416/180)))*(tan(ALFA_1*(3.1416/180)));
INT_VAR5 = (X1*X1-49*D0*D0)*INT_VAR4;
INT_VAR6 = 2*V0*K*D0*(1-exp(-INT_VAR4));

time_2 = INT_VAR5/INT_VAR6;

INT_VAR7 = (tan(THEATA_1*(3.1416/180)))*(tan((3.1416/180)*(THEATA_1-ALFA_1)));
D_01 = (D0/2)+(1+INT_VAR7)*s*(tan(ALFA_1*(3.1416/180)));

INT_VAR8 = (tan(THEATA_1*(3.1416/180)))*(tan((3.1416/180)*(THEATA_1+ALFA_1)));
D_02 = (D0/2)+(1+INT_VAR7)*s*(tan(ALFA_1*(3.1416/180)));

INT_VAR9 = (((sin(THEATA_1*(3.1416/180)))*(sin(THEATA_1*(3.1416/180))))+
(0.81*(cos(THEATA_1*(3.1416/180)))*(cos(THEATA_1*(3.1416/180)))));
INT_VAR10 = sqrt(INT_VAR9);
INT_VAR11 = 0.9*(cos(THEATA_1*(3.1416/180)));
X2 = s*(INT_VAR10)/(INT_VAR11);

VH_21 = V0*K*D0*(1-exp(-INT_VAR4))/(X1*INT_VAR4);
INT_VAR12 = 2*VH_21*K*(D_01+D_02)*(1-exp(INT_VAR4))
*INT_VAR10;
INT_VAR13 = X2*X2*INT_VAR4;

time_3 = INT_VAR13/INT_VAR12;

time_total_calculated=time_1+time_2+time_3;
INT_VAR14=X1+X2;
Diff=Xf- INT_VAR14;
```

```
                    if(time_total_calculated<0.001 && time_total_calculated>0.0009)
                    {
                              if(Diff<0.00001 && Diff>0)
                              {
                              printf("\n %f \t %f \t %f \t %f \t %f \t %f \t %f \t %f \t
%f ",time_total_calculated,Xf,INT_VAR14,Diff,X1,X2,time_1,time_2,time_3);
                              fprintf(fp,"\n %f \t %f \t %f \t %f \t %f \t %f \t %f \t %f
\t %f ",time_total_calculated,Xf,INT_VAR14,Diff,X1,X2,time_1,time_2,time_3);
                              }
                    }
          }


     }
}
```

(2) Volume

```
# include<stdio.h>
# include<math.h>
# include<conio.h>

main( )
{

float T_H2_IN,A1,A2,T_H2_OUT,m1,Xf,X1,X2,D0,y1,M_H2,V0,K,Re;
float  time_total,time_total_calculated,T_calu,time_1,time_2,time_3;

float
INT_VAR1,INT_VAR2,INT_VAR3,INT_VAR4,INT_VAR5,INT_VAR6,INT_VAR7,INT_VAR8,INT_VAR9,INT
_VAR10;float
INT_VAR11,INT_VAR12,INT_VAR13,INT_VAR14,INT_VAR15,INT_VAR16,INT_VAR17,
INT_VAR18,INT_VAR19,INT_VAR20,Diff;

float  tolerance,tolerance2;

float M_COAL,V_H2_I,CP_H2_AV,CP_COAL;
float a,b,x,Miu,sp_vol,Mass_H2,row_H2,Miu_by_row,V_H2r,ds;
float VH_21,D_01,D_02,s,ALFA_1,THEATA_1,w,h,v;
```

```
float Miu_c, sp_vol_c,row_H2_c,Miu_by_row_c,D_03,D_04,e1;
printf("Enter the value of input temp(T_H2_IN) = ");
scanf("%f ",&T_H2_IN);
printf("\n Enter the value of Xf = ");
scanf("%f",&Xf);
printf("\n Enter the value of X1 = ");
scanf("%f",&X1);
A1 = T_H2_IN+940;
A2 = T_H2_IN+21.5;
//ALFA_1=3;
e1=0.9;
THEATA_1 = 4;
T_H2_OUT = 723.0;
M_COAL = 0.01157;
CP_H2_AV = (0.0179*(T_H2_IN+ T_H2_OUT)*1000/2);
CP_COAL = 1008.0;
M_H2 = ((M_COAL*CP_COAL*425)/(CP_H2_AV*(T_H2_IN- T_H2_OUT)));

printf("Enter the value of V_H2_I(m/s)= ");
scanf("%f",&V_H2_I);

V0=(M_H2*V_H2_I)/(M_H2+M_COAL);

//Miu_by_row at the inlet temp
Miu=(0.000000015*T_H2_IN+0.000005031);
sp_vol = (0.0006*T_H2_IN+0.0165);
Mass_H2 = 2;
row_H2 = Mass_H2/sp_vol;
Miu_by_row = Miu/row_H2;

// Dia of inlet jet
D0 = sqrt((M_H2*sp_vol*4)/(V_H2_I*3.1416*2));
Re = (D0*V0/Miu_by_row);
K = 1.41*pow(Re,0.135);
time_total = 0.001;
y1 = 7*D0;
```

```
m1 = (T_H2_OUT-T_H2_IN)/(Xf);
        // defining intermidiatre variable
INT_VAR1 = (306.5*D0*D0*D0*(pow(A1,0.27))/(M_H2*(pow(A2,0.73))));
INT_VAR2 = (2.48-123.113*D0*((0.27*m1/A1)-(0.73*m1/A2)));
INT_VAR3 = 0.197*m1*m1*(4607.196*D0*D0)/(A1*A2);
time_1 = INT_VAR1*(INT_VAR2+INT_VAR3);
        // T = a+b.x
        a = T_H2_IN;
        b = (T_H2_OUT-T_H2_IN)/Xf;
        T_calu = a+b*7*D0;
        Miu_c = (0.000000015*T_calu+0.000005031);
        sp_vol_c = (0.0006*T_calu+0.0165);
        row_H2_c = Mass_H2/sp_vol_c;
        Miu_by_row_c = Miu_c/row_H2_c;
        //code changed
        ALFA_1 = 0.238*(pow(Miu_by_row_c,0.135))*(180/3.1416);
        s = X1*(cos(THEATA_1*(3.1416/180)));

INT_VAR4 = 92.1*(tan(ALFA_1*(3.1416/180)))*(tan(ALFA_1*(3.1416/180)));
        INT_VAR5 = (X1*X1-49*D0*D0)*INT_VAR4;
        INT_VAR6 = 2*V0*K*D0*(1-exp(-INT_VAR4));

        time_2 = INT_VAR5/INT_VAR6;

INT_VAR7 = (tan(THEATA_1*(3.1416/180)))*(tan((3.1416/180)*(THEATA_1-ALFA_1)));
        D_01 = (D0/2)+(1+INT_VAR7)*s*(tan(ALFA_1*(3.1416/180)));

INT_VAR8 = (tan(THEATA_1*(3.1416/180)))*(tan((3.1416/180)*(THEATA_1+ALFA_1)));
        D_02 = (D0/2)+(1+INT_VAR7)*s*(tan(ALFA_1*(3.1416/180)));

INT_VAR9 = (((sin(THEATA_1*(3.1416/180)))*(sin(THEATA_1*(3.1416/180))))+(0.81*
(cos(THEATA_1*(3.1416/180)))*(cos(THEATA_1*(3.1416/180)))));
        INT_VAR10 = sqrt(INT_VAR9);
        INT_VAR11 = 0.9*(cos(THEATA_1*(3.1416/180)));
        X2 = s*(INT_VAR10)/(INT_VAR11);

        VH_21 = V0*K*D0*(1-exp(-INT_VAR4))/(X1*INT_VAR4);
        INT_VAR12 = 2*VH_21*K*(D_01+D_02)*(1-exp(-INT_VAR4))
*INT_VAR10;
```

```
INT_VAR13 = X2*X2*INT_VAR4;

time_3 = INT_VAR13/INT_VAR12;

time_total_calculated = time_1+time_2+time_3;

INT_VAR14=X1+X2;
Diff= Xf- INT_VAR14;
//*******************************************************************
INT_VAR15 = (3.1416/180);
//(tan(THEATA_2) = INT_VAR16
INT_VAR16 = (tan(THEATA_1*(3.1416/180)))/e1;
//(tan(THEATA_2-ALFA_1) = INT_VAR17
INT_VAR17 = (INT_VAR16-(tan(ALFA_1*(3.1416/180))))/(1+INT_VAR16*(tan
(ALFA_1*(3.1416/180))));
//(tan(THEATA_2+ALFA_1) = INT_VAR18
INT_VAR18 = (INT_VAR16+(tan(ALFA_1*(3.1416/180))))/(1–INT_VAR16*(tan(ALFA_1*
(3.1416/180))));

D_03=
D_01+(1+(tan(THEATA_1*(3.1416/180))/e1)*INT_VAR17)*s*(tan(ALFA_1*(3.1416/180)));
D_04=
D_02+(1+(tan(THEATA_1*(3.1416/180))/e1)*INT_VAR18)*s*(tan(ALFA_1*(3.1416/180)));
w= D_03+D_04+0.02;
h=(5*3.14285*D0*D0/(8*w))+(D0/2)+s*((tan(THEATA_1*(3.1416/180)))+((tan(THEATA_1*
(3.1416/180)))/e1))+D_04+0.01;
v = s*w*h*pow(10,6);
printf("\n T_H2_IN = %f ",T_H2_IN);
printf("\n D0(mm) = %f ",D0*1000);
printf("\n D_01(mm) = %f ",D_01*1000);
printf("\n D_02(mm) = %f ",D_02*1000);
printf("\n D_03(mm) = %f ",D_03*1000);
printf("\n D_04(mm) = %f ",D_04*1000);
printf("\n vol(cc) = %f ",v);
}
```

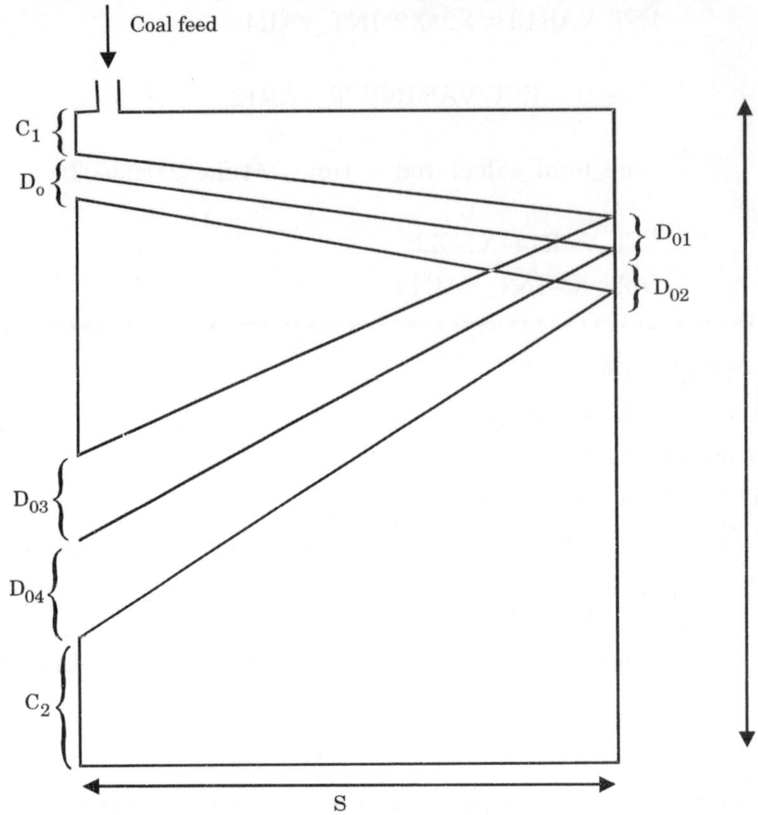

Figure 2.2: Jet dimensions

Table 2.1: Volume of preheater vs initial velocity of hydrogen

Initial Temperature of Hydrogen (500°C)

Velocity of hydrogen (m/s)	Volume of preheater (cc)
125	24.40
150	24.88
200	26.00
250	27.19
300	28.37

Table 2.2: Volume of preheater vs initial temperature of hydrogen

Initial Velocity of Hydrogen (300 m/s)

Initial Temperature of hydrogen (°C)	Volume of preheater (cc)
500	28.37
550	12.65
600	7.89
650	5.68

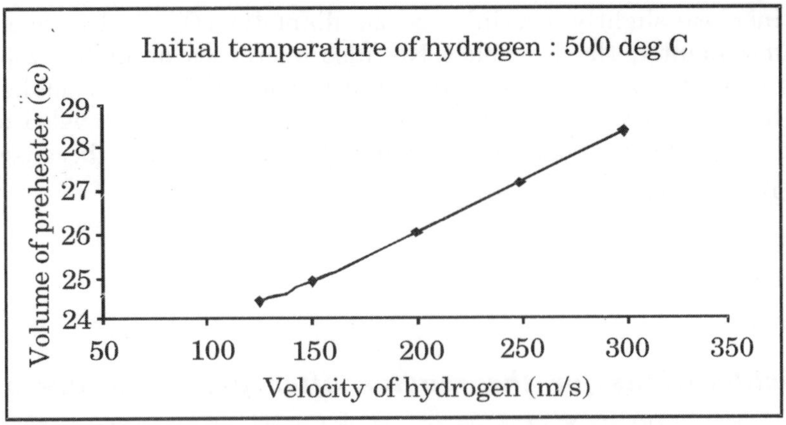

Figure 2.3: Volume of preheater vs Initial velocity of hydrogen

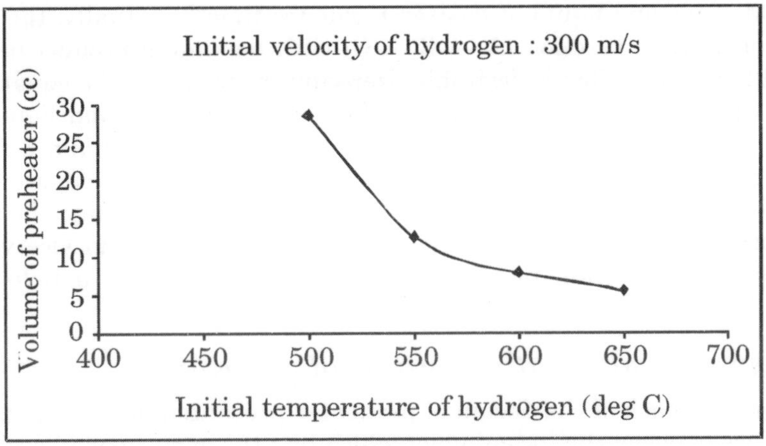

Figure 2.4: Volume of preheater vs Initial temperature of hydrogen

2.2 RESULTS AND DISCUSSION (Refer Tables 1 and 2)

From Figures 2.3 and 2.4 one can conclude that with increase in the inlet velocity of hydrogen the volume of the preheater increases for a given inlet temperature of hydrogen. Also, with increase in the initial temperature of hydrogen, the inlet velocity of hydrogen remaining same, the volume of the preheater decreases.

The above results may be explained by the fact that due to fixed total residence time of 1 millisecond in the multipass preheater, with the increase in the velocity of hydrogen at a given inlet temperature of hydrogen the following results :

1. Inlet nozzle diameter D_0 decreases. So, D_{01}, D_{02}, D_{03} & D_{04} decrease.
2. But 's' increases due to higher velocity of hydrogen.

Thus, 'w' decreases, 's' increases and 'h' happens to increase also due to large contribution of s although D_0 and D_{04} are less. Hence, the preheater volume increases with increasing inlet velocity of hydrogen at a particular initial temperature of hydrogen. Also, with increase in inlet temperature of hydrogen, mass flow rate of hydrogen decreases and although specific volume may increase slightly, the inlet nozzle diameter (D_o) for hydrogen flow eventually decreases. Subsequently, the Reynold's No. also decreases at higher temperature. So 'K' decreases and hence from equations (4) and (5) it is evident that with smaller 'D_o' and smaller 'K', x_1 and x_2 have to decrease for same't' meaning that distance traveled by the gas stream is less. Hence the preheater volume is less at higher temperature, because of the following:

1. D_{01}, D_{02}, D_{03} and D_{04} decrease due to lower D_o
2. w decreases
3. s decreases
4. h decreases

Design considerations for the reactor after preheater are as follows:

From the outlet of the preheater one has to reduce the cross-section of the pipeline reactor. This is because one has to achieve turbulent flow (close to plug flow) because that will lead to maximum yield of tar and liquid at desired residence time[(8)]. Actually, the reaction sequence is similar to consecutive reaction (A $\rightarrow$ B $\rightarrow$ C) with maximum product being 'B' that is tar & liquid[(8)] and hence plug flow is desirable. Deriving theoretically the equations for prediction of length and diameter of reactor for the yield of maximum tar and liquid (numerically) is rather complicated. One has to do experiments to determine optimum diameter and length of the reaction zone after preheater. Length will depend on desired residence time for maximum yield of 'B' and optimum diameter is to be determined from economics i.e. smaller diameter means more pressure drop but better yield (close to plug flow) and vice versa. Refer to the rate equations for formation of tar and liquid in a previous article[(8)] by the author (A. Sirkar).

Modifications and alternative design of preheater

One major drawback of the design of preheater presented here is the fact that there will occur interference between the incoming jet which hits the opposite wall and the bounced back jet which returns. To minimize this problem one can design the opposite wall of the preheater to be at an angle as shown in Figure 2.5. By this design the angle of reflectance will increase thus, lessening the degree of interference between the two jets.

An alternative design is shown in Figure 2.6. The advantage of this design is that the jet does not hit the wall, thus lessening the possibility of plastic coal sticking to solid surface except may be at the periphery of the bend which could be avoided by carrying out experiments with variation in forward and backward jet travel distance that will make sure that plastic zone is not where the opposite wall or the bend exists. In absence of any correlation for rate of heat transfer to coal particles by the free turbulent jet, assume linear temperature profile (see item (*ii*) under design basis) for coal particles and make sure that the plastic region i.e. 325 to 350^{0}C temperature is reached just after the jet hits the opposite wall in order to avoid plastic coal sticking to solid wall and jeoparadising the liquefaction process. Note that linear temperature profile is a conservative design approach because plastic zone temperature may be achieved much earlier because of possibly greater turbulence at the beginning of forward path due to higher jet velocity meaning higher rate of heat transfer. By varying the length of forward and backward path (in particular, increase in length of the forward path) not only one can avoid plastic zone at the wall but improved yield of tar and liquid can be obtained. This is because higher jet velocity can be maintained over a longer length meaning higher turbulence and shorter residence time. Shorter residence time means that coal stays shorter and thus less of it gets converted to tar, liquid and gas during preheating as because any tar and liquid that is produced during preheating, less of it would get converted to gas due to shorter residence time. Note that in the backward path, jet velocity gets reduced after hitting the wall which is why it is better to have a shorter backward path. Note that higher turbulence due to higher jet velocity also means quicker heating due to higher rate of heat transfer.

In the alternative design, average velocity of the jet does not decrease due to passage through the bend unlike the case when jet bounces back thus, insuring greater degree of turbulence within the jet due to higher average jet velocity.

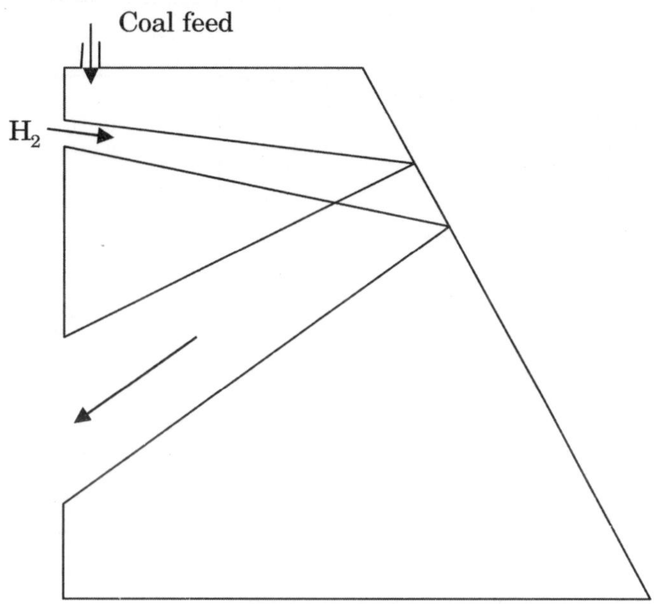

Figure 2.5: Modified preheater design

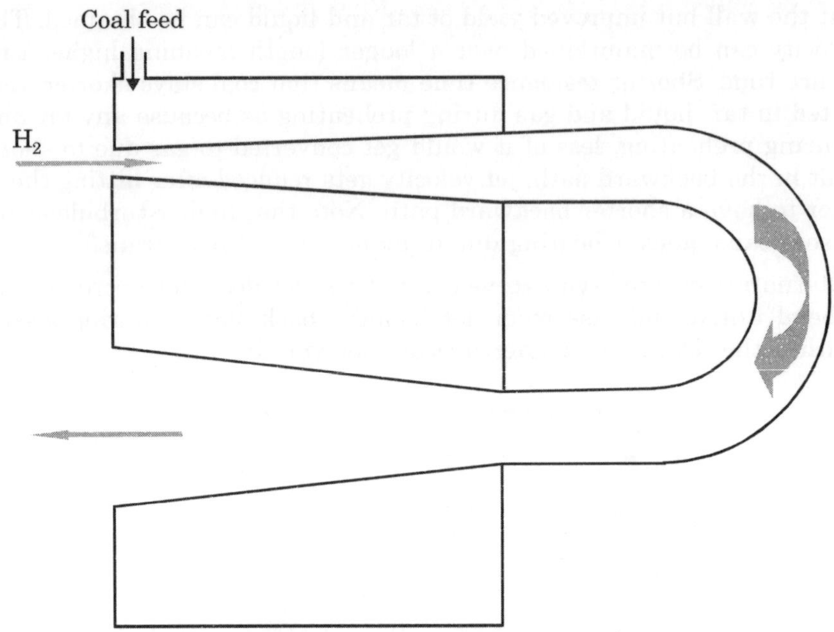

Figure 2.6: Alternative preheater design

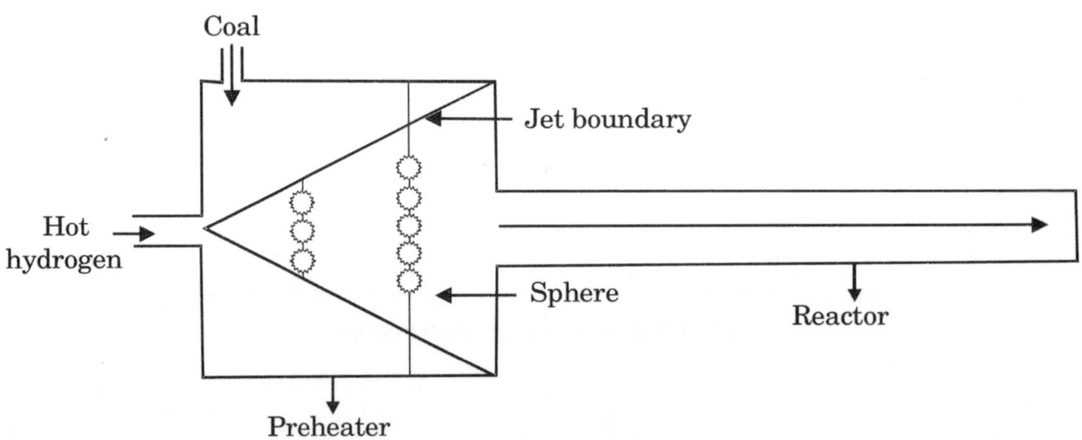

Figure 2.7

2.3 HEAT TRANSFER ASPECTS OF PREHEATER DESIGN

So far fluid dynamics aspects of preheater design have been considered. It has been assumed that within .001 second residence time coal will be preheated to 450˚C. Since no heat transfer coefficient correlation is available for free jet entraining solid particles, fixed bed heat transfer coefficient correlation[9] has been used to check whether coal does actually get heated to 450˚C within .001 second residence time. The results of calculations are given below.

Assumption

1. Coal remains as solid particle till it reaches the temperature of 450°C.

Since coal passes through plastic region, heat transfer rate in this region may be lower due to large fluid mass than consideration of individual solid particle, and hence if heat supplied is more than heat required using fixed bed correlation then the design is O.K. This is shown below in the example.

Example

$V_{H_{2_{in}}} = 300$ m/s

$T_{H_{2_{in}}} = 873°$ K

$t_1 + t_2 = 0.000368$ sec

$t_3 = 0.000632$ sec

$d_p = 7.4 \times 10^{-5}$m

$G = 80\,\Re\text{g s}^{-1}\text{m}^{-2}$

Note that G changes along the jet path. It is highest at the inlet and decreases along the jet path from entrance to exit. For calculation of Reynolds number the value of G has been taken at the end of forward jet path.

Using the above input data, heat supplied and heat required for heating coal to the desired temperature (450°C) have been calculated as follows, assuming 5°C temperature difference between hydrogen and coal at the end of backward jet path.

Heat supplied = 7,538 J/s

Heat required = 4,957 J/s

Therefore, 0.001 s contact time is sufficient to preheat coal to the desired temperature.

A better estimate of heat supplied could be made by calculating h_{av} by graphical integration because G as well as C_{p_g}, μ_{H_2} and k change from entrance to exit. h_c is given by the following correlation[9] :

$$\frac{h_c d_p}{k} = 0.65\left(\frac{d_p G}{\mu_{H_2}}\right)^{0.6}\left(\frac{C_{p_g}\mu_{H_2}}{k}\right)^{0.33}$$

Coal Residence Time:

It is expected that Mean Residence Time will be higher than Hydraulic Residence Time (0.001) – to be checked through calculation. If it is higher then coal stays longer in the preheater resulting in reactions in addition to preheating as elaborated earlier. With reactions starting earlier liquid yield decreases because it gets gasified. To improve the yield of liquid cut back the hydraulic residence time from 0.001 sec to lower than this in which case Mean Residence Time will be lower than before thus, coal stays shorter in preheater and liquid going to gas will be at a minimum. Actual hydraulic residence time to be used depends on heat transfer requirement guaranting that coal will be heated to desired operating temperature by that time. From the above calculations it appears that in less than 0.001 sec coal gets heated to 450°C assuming packed bed heat transfer coefficient. Actually, heat transfer rate will be higher than packed bed one because of free turbulent jet. So one can safely take less than 0.001 sec as hydraulic residence time and calculate preheater dimensions accordingly.

2.4 VERIFICATION OF THE DESIGN CONCEPTS OF PREHEATER–CUM–REACTOR

Experiments have been carried out to verify the concepts and it has been found that yield of tar and liquid is maximum with the alternative design (Fig. 2.6) and minimum with the box type design (Fig. 2.2) for which the equations are given in the text for the free jet turbulence concept. However, the design concept where spheres[13] are used inside the preheater (Figure 2.7) for creating turbulence (due to boundary layer separation) yield of tar and liquid is much higher. Mixing is much better in this design and residence time is close to .001 sec. In case of free jet turbulence the jet has a velocity profile with minimum velocity at the periphery of the jet such that it takes longer time for coal to pass through the preheater. If coal stays longer in the preheater it undergoes reaction to tar, liquid and gas and tar and liquid undergoes gasification thus lowering tar and liquid yield. Besides, incoming jet and reflected jet undergoes collision inside the preheater for box type preheater and may follow a zig-zag path before coming out of the outlet which is also the reason for less liquid and tar yield because of long residence time that is greater than 0.001 second. In order to improve the yield with jet preheater use the alternative design (Figure 2.6) because no collision occurs here which otherwise usually results in longer residence time in preheater.

2.5 NAPHTHA PREHEATING

For naphtha preheating one can use a preheater in which free jet turbulence is coupled with spheres which results in additional mixing behind the spheres due to boundary layer separation. Only difference between coal preheater and Naphtha preheater is that in case of coal it is heating solid by gas whereas in case of naphtha it is solid to gas heat transfer.

The heating time in this case is 0.01 second, on the basis of which the length of the preheater can be determined using equations for average jet velocity for the two zones namely, the zone where flow is not fully developed followed by the fully developed region. Equations for average velocity of jet and jet residence time similar to coal preheater could be developed for naphtha preheater provided one knows the temperature profile of naphtha vapour which can be determined experimentally. The profile is not linear due to the fact that the region where spheres are located, better mixing causes abrupt change in profile. If one uses box type preheater then one could assume linear profile like coal preheater of similar type.

2.6 REACTOR FOR LIQUID PHASE HYDROGENATION OF COAL

Here we shall concentrate on H-Coal Process Reactor which is an ebullated bed reactor (Fig. 2.8). Scale-up of such reactor is the main issue here. Different scale-up techniques may be adopted to duplicate the pilot plant yield in a commercial reactor.

(*a*) One scale-up technique is to use vertical baffles[10] (Figure 2.9) in order to maintain same L/D ratio as in pilot plant reactor such that the hydrodynamics is similar *i.e.*

$$\left(\frac{L}{D}\right)_{\text{Pilot Plant Reactor}} = \left(\frac{L}{D_{eq}}\right)_{\text{Commercial plant}}$$

The vertical baffles divide the large reactor into smaller reactors, each of those representing pilot-plant reactor in principle but not exact scale-up because the cross-section is non-circular having different fluidization characteristics, that affects yield of products due to lateral mixing currents, meaning production of more gaseous substance due to back- mixing.

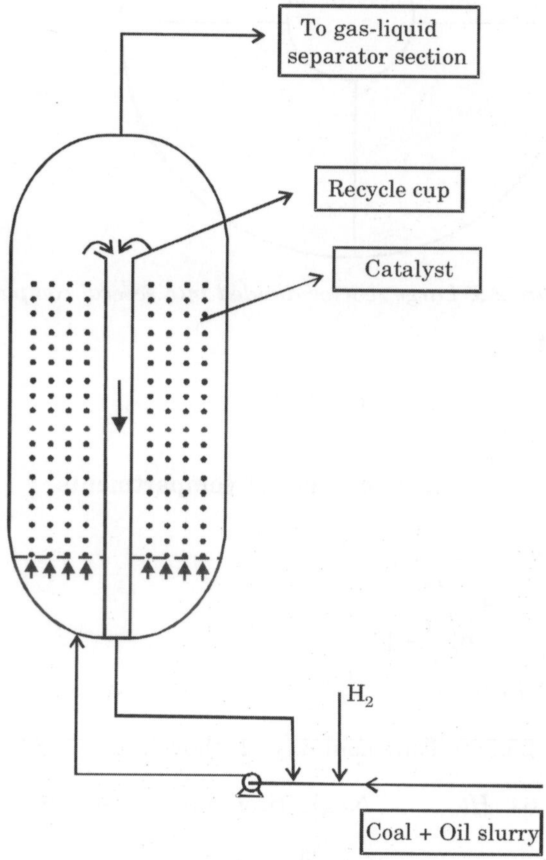

Figure 2.8: Ebullated bed reactor for coal liquefaction

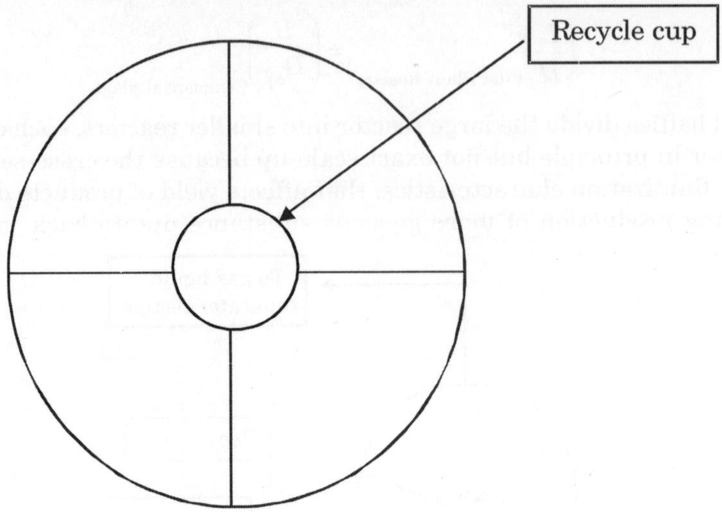

Figure 2.9: Large reactor divided into several compartments

$$4R_H = D_{eq} = \frac{4\frac{A_{cp}}{n}}{\frac{\pi D_{cp}}{n} + D_{cp}} \qquad A_{cp} = \frac{\pi}{4} D_{cp}^2$$

n – no. of compartments

$$= \frac{\cancel{4} \times \frac{\pi}{\cancel{4}} \frac{D_{cp}^{\cancel{2}} D_{cp}}{n}}{\cancel{D}_{cp}\left(\frac{\pi}{n}+1\right)} = \pi \frac{D_{cp}}{n\left(\frac{\pi}{n}+1\right)}$$

Basis: Commercial Reactor

SV is 36.2 $\frac{\text{lb}}{\text{hrft}^3}$, 25,000 Tons coal/day, 1 short ton = 2000 lbs, 8 reactors

D_{cp} = 12′ *ID*, *L* = 64′ *Ht* Neglecting the recycle cup area

$$\left(\frac{L}{D}\right)_{pp} = 75 = \frac{\cancel{64}^{16}}{\pi \times \frac{\cancel{12}^{3}}{n\left(\frac{\pi}{n}+1\right)}} = \frac{16n\left(\frac{\pi}{n}+1\right)}{3\pi}$$

$$n\left(\frac{\pi}{n}+1\right) = \frac{3\pi \times 75}{16} \qquad \pi + n = \frac{3\pi \times 75}{16}$$

$$\therefore \quad n = \frac{3\pi \times 75}{16} - \pi$$

$$= 3.14\left(\frac{225}{16} - 1\right)$$

$\sqcup$ 41 compartments

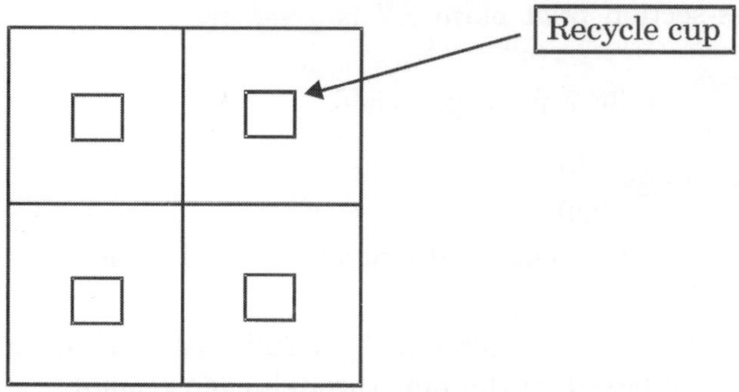

Figure 2.10: Square cross-section large reactor divided into several square compartments

(b) If we carry out experiment in pilot plant reactor which is of square cross-section and scale-up by dividing the square cross section large reactor into smaller squares then one can duplicate the pilot plant yield in commercial reactor effectively[11](Figure 2.10).

Basis: 25,000 Ton coal/day SV = $28 \dfrac{\text{lb}}{(\text{hr})(\text{ft}^3)}$

Conditions: 850°F, 3000 psig

Total gas at conditions – 126.6 ft³/sec

If we use a large reactor of 16 ft square then

$$n_R \times (16)^2 \times v_g = 126.6$$

$$n_R = \frac{126.6}{256 \times 0.18} = 2.74 \simeq 3$$

where v_g = 0.18 fps (gas velocity) for 80% bed expansion with liquid velocity of 0.105 fps

$$L = \frac{57551 \times 36.2}{3 \times 256 \times 28} = 93.5$$

$$\left(\frac{L}{D}\right)_{cp} = 75$$

16 ft square can be reduced to 256 squares each of side 1 ft.

$$\therefore \quad \frac{L}{D_{cp}} = 93.5 >> 75$$

Therefore, one has to use a larger dia reactor. Let it be 20′ square.

$$\text{Then, } n_R = \frac{126.6}{(20)^2 \times .18} \quad 1.76 \simeq 2$$

$$L = \frac{57551 \times 36.2}{28 \times 2 \times (20)^2} = 93$$

20′ square →10′ → 5′ → 2.5′ → 1.25′ each side

$$\frac{93}{1.25} = \frac{L}{D} = 74.4 \simeq 75 \text{ (closest)}$$

In square cross-section pilot plant SV is given by

$$\text{SV} = \frac{\pi a^2}{4 \times a^2} \times 36.2 \text{ for same yield}$$

$$= 28.42 \simeq 28 \frac{\text{lb}}{\text{hrft}^3}$$

where a is diameter of circular pilot plant reactor and is also each side of the square pilot plant reactor.

Note that the smaller cross-section (1.25′ × 1.25′) represents each reactor with recycle cup of appropriate dimension based on the flow through each reactor.

For the 20′ square large reactor there are 256 smaller reactors.

(*c*) Another scale-up technique is to use a large number of pilot plant dimension reactors inside a single shell. Thus yield of product will be same in commercial reactor. Only disadvantage with this design is that due to circular cross-section of the individual reactors there will exist some void space inside the shell which is of no use and hence the outer shell diameter will be large. However, in comparison to square cross-section here SV is higher because of no dead space as in square reactor and hence active volume of reactor space required is less and fabrication cost is less also. But, in square cross-section there exists common walls between the smaller squares and hence, wall thickness can be reduced (requiring less material) due to same pressure acting on both sides of the wall.

2.7 GAS–SOLID CATALYTIC REACTOR

In any gas-solid catalytic reaction there are usually several resistances in series for getting products of reaction as follows:

1. Diffusion of reacting gases from bulk to the vicinity of catalyst
2. Adsorption of reactants to active catalysts sites
3. Surface reaction at catalyst surface
4. Desorption of products of reaction from catalyst surface to the vicinity of catalyst
5. Diffusion of products of reaction from vicinity of catalyst surface to bulk.

Diffusional resistances (step 1 and 5) can be reduced by increasing the linear velocity of the gas but this results in increased pressure drop. An alternative and better technique would be use of non-conventional catalyst shapes and vibrating the reactor as follows:

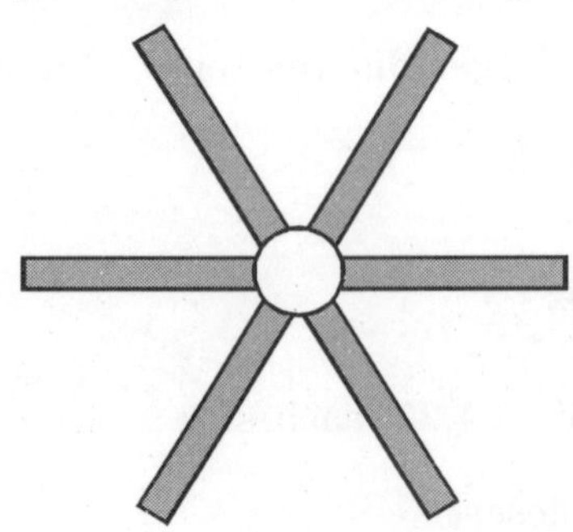

Figure 2.11

Let us consider a pipeline reactor (square duct) containing catalyst particles in the shape of a star made up of curved cross-sectional arms of equal length and width (see Figure 2.11). The pipeline is half filled up with catalyst particles, the reason for which is given in chapter 3 under gas-gas mixing. The pipeline is freely supported at two ends and due to impact particles move up and down. During upward and downward movement of the particles gases are displaced in the form of waves due to compressive action of the particles on the gas. Due to curved cross-section of the arms, waves from adjacent arms move in a curvilinear path and undergo collision in the space between two adjacent arms, below and above the particles due to flimsy surface of the waves. The waves penetrate each other due to low viscosity in addition to their momentum. Due to resistance offered by one wave on the other, when they encounter, there will be a tendency of the waves to move backwards (taking the path of least resistance) when they collide, thus, generating a circulating motion. This results in mixing within the waves. The waves have a concentration gradient in them due to diffusion and this gradient gets destroyed due to the mixing phenomena described above. Thus, mass transfer rates are enhanced and in addition the adsorption rate increases due to higher concentration of reactants at the catalyst surface. Thus, the overall rate of reaction gets enhanced. However, desorption rate of SO_3 will decrease because of the mixing phenomena, since the concentration of SO_3 decreases at the arms.

Example: $SO_2 + ½O_2 \leftrightarrow SO_3$

Mathematical model of the conventional reactor (packed bed) for the above reaction is given in Rase's book (Vol. 2)[12]. Here, only the relevant heat transfer and mass transfer rate equations for the above mentioned modified reactor (vibrated bed) using non-conventional catalyst shape are given.

Mass Transfer Equation

$$\frac{(V_v + V_t) \times N_c \times A_c \times W_p \times dL}{y_f \times \rho_{gL}} \times \left(\frac{(C_{SO_{2_i}} + C_{SO_{2_b}})}{2} - C^*_{SO_{2_i}} \right) = r_{SO_2} \times A_c \times dL \times \rho_b \quad ...(10)$$

where $C_{SO_{2_i}} < C^*_{SO_{2_i}} \le \left(\frac{C_{SO_{2_i}} + C_{SO_{2_b}}}{2} \right)$ due to two opposing factors namely, mixing and diverging arms.

Heat Transfer Equation

$$\frac{(V_v + V_t) \times N_c \times W_p \times C_{Pg} \left(t_s - \frac{t_s + t_b}{2} \right)}{y_f} = r_{SO_2} \times \Delta H_r \times \rho_b \quad ...(11)$$

Rate of transfer of heat from catalyst surface to bulk = Heat evolved due to reaction. y_f, the wave thickness depends on the thrust of the particles on the gases and hence may be correlated to W_p, V_v and V_t as follows:

$$y_f = a\,(W_p\, V_v)^b\, (W_p\, V_t)^c \quad ...(12)$$

Note that higher the thrust, larger will be y_f

For solving the system of equations given in Rase's[12] book together with the above equations for mass and heat transfer for determining the length of the reactor one has to choose particle size and V_v. Particle size fixes W_p and V_t.

It is assumed that when catalyst hits the top surface its velocity is zero so that it attains terminal velocity while falling back. Actually, one has to carry out experiment, get the conversion of SO_2 for a given length of reactor and then solve the equations and back calculate y_f which is then to be correlated to W_p, V_v, and V_t. After one develops the correlation for y_f one can choose proper values of W_p, V_v and V_t for maximum conversion in a given length of the reactor.

2.8 DEVELOPMENT OF MASS TRANSFER EQUATION

1. Momentum Balance Equation:

$$N_c \times A_c \times dL \times W_p \times V_v = N_c \times A_c \times dL \times A_p \times y_f \times V'_v \times \rho_{gL} \quad ...(13)$$

$$N_c \times A_c \times dL \times W_p \times V_t = N_c \times A_c \times dL \times A_p \times y_f \times V'_w \times \rho_{gL} \quad ...(14)$$

2. Mass Transfer Equation:

$$N_c \times A_c \times dL \times A_p \times (V'_w + V'_v) \times \left(\frac{C_{SO_{2_i}} + C_{SO_{2_b}}}{2}\right) - C^*_{SO_{2_i}} = r_{SO_2} \times A_c \times dL \times \rho_b \quad ...(15)$$

or

$$\frac{N_c \times \cancel{A_c} \times \cancel{dL} \times W_p \times (V_t + V_v)}{y_f \times \rho_{gL}} \left(\frac{C_{SO_{2_c}} + C_{SO_{2_b}}}{2}\right) - C^*_{SO_{2_i}} = r_{SO_2} \times \cancel{A_c} \times \cancel{dL} \times \rho_b \quad ...(16)$$

Rate of transfer of SO_2 from bulk to the vicinity of catalyst = rate of reaction of SO_2 to SO_3.

In a similar fashion the Heat Transfer Equation could be developed.

Equation (10) is valid for small particles and near the central region of large catalyst particle only, because at a distance away from the centre, concentration of SO_2 in the vicinity of the arms decreases due to diffusion of SO_2 from the mixed wave region to catalyst surface, since, mixing of waves occurs at a distance from the arms. Similar argument holds for equation (11) regarding temperature.

2.9 VERIFICATION OF THE DESIGN CONCEPT FOR GAS-SOLID CATALYTIC REACTOR

Experiments have been carried out for various gas-solid catalytic reactions using the reactor design suggested in the text and it has been found that length of reactor required for a given conversion is significantly less than the conventional packed bed reactor (using either cylindrical or spherical catalyst particles). Although, the vibrating mechanism (coupled with power consumption) and particular shape of catalyst (star shaped) will incur additional cost compared to simple packed bed reactor using conventional shaped catalyst, it has been found that there is a net cost benefit due to significant reduction in length of reactor and catalyst volume. Note that the reactor is only half filled with non-conventional shaped catalyst particles (refer to chapter 3 of Section II).

NOMENCLATURE

A_{cp}	–	cross-sectional area of commercial reactor in ft^2
A_c	–	cross-sectional area of gas-solid catalytic reactor in m^2
A_p	–	surface area per particle in m^2
$C_{SO_{2_i}}$	–	concentration of SO_2 at catalyst surface in $\frac{\text{kgmole}}{m^3}$
$C_{SO_{2_b}}$	–	concentration of SO_2 in the bulk of the gas in $\frac{\text{kgmole}}{m^3}$
C_{p_g}	–	specific heat of gas in $\left(\frac{\text{kcal}}{\text{kgmole}(°K)}\right)$
$C_{p_{coal}}$	–	specific heat of coal in Joule/(kg) (°K)
$C_{pH_{2av}}$	–	average specific heat of hydrogen in Joule/(k_g) (°K)
C_1	–	clearance above H_2 jet inlet, meter
C_2	–	clearance below outgoing jet, meter
D	–	diameter of pilot plant reactor in feet
D_{eq}	–	equivalent diameter of reactor in feet
D_{cp}	–	diameter of commercial plant reactor in feet
D_o	–	inlet diameter of nozzle through which hydrogen enters the preheater, meter
d_p	–	coal particle size in meter
e	–	coefficient of elasticity
G	–	mass velocity of hydrogen, $\frac{\text{kg}}{(m^2)(\text{sec})}$
h	–	height of preheater, meter
h_c	–	heat transfer coefficient in $\frac{\text{kcal}}{(m^2)(\text{sec})(°C)}$
ΔH_r	–	heat of reaction in kcal/kgmole
k	–	thermal conductivity of hydrogen in $\frac{\text{kcal}}{(m)(\text{sec})(°K)}$
L	–	length of gas-solid catalytic reactor in meter
m_{H_2}	–	inlet hydrogen flow rate, kg/sec
n	–	number of compartments
n_R	–	number of reactors
N_c	–	concentration of particles in number per unit volume (m^3) of reactor
r	–	radial distance from jet centerline, meter
r_{SO_2}	–	rate of reaction of SO_2 in kgmole SO_2/(kg cal) (sec)

s	–	length of preheater, meter
$T_{H_{2in}}$	–	inlet temperature of hydrogen, °K
$T_{H_{2out}}$	–	outlet temperature of hydrogen, °K
t_s	–	temperature of gas at catalyst surface in °C
t_g	–	temperature of gas in the bulk in °C
$\overline{u_x}$	–	average velocity of jet in the region where flow is not fully developed, meter/sec
$V_{H_{2i}}$	–	inlet velocity of hydrogen, meter/sec
$V_{H_{21}}$	–	average incident jet velocity, meter/sec
$V_{H_{22}}$	–	average velocity of the bounced jet, meter/sec
V_o	–	velocity of coal and hydrogen mixture at the entrance to preheater, meter/sec
V_c	–	centerline velocity of jet, meter/sec
V_r	–	horizontal velocity of jet at any radius r, meter/sec
V_v	–	vibration velocity in meter / sec
V_t	–	terminal velocity of particle in meter/sec
V'_v	–	velocity of wave during upward movement of particle in meter/sec
V'_w	–	velocity of wave during downward movement of particle in meter/sec
w	–	width of preheater, meter
W_p	–	weigt per particle in kg
x	–	distance along jet centerline, meter
x_1	–	forward jet path distance along jet centerline, meter
x_2	–	backward jet path distance along jet centerline, meter
y_f	–	wave front thickness in meter

Greek letters

α	–	half of solid angle of the jet
α_1	–	half of solid angle of the jet at $x_1 \geq 7D_o$
θ_1	–	angle of incidence of the jet w.r.t. horizontal
θ_2	–	angle of reflectance of the jet w.r.t. horizontal
ρ_b	–	bulk density of catalyst in kg/m^3
ρ_{gL}	–	density of gas in kg/m^3
ρ_{H_2}	–	density of hydrogen, kg/m^3
μ_{H_2}	–	viscosity of hydrogen, $\frac{kg}{(m)(sec)}$

REFERENCES

1. Sirkar, A., Indian Patent 208758 (2007).
2. Personal communication with Dr. W.C. Shroeder (USA).
3. Chakravorty, J.G. and Ghosh, P.R., *Analytical Dynamics,* Moulik Library Publication, 1997, Kolkata.
4. Rushton, J.H., 'The axial velocity of a submerged axially symmetrical fluid jet' AIChE J Vol. 26, 6, 1038 (1980).
5. Perry, R.H., Green, D.W., Perry's Chemical Engineers' Handbook, Seventh edition, McGraw Hill (1997), USA..
6. Hill, B.J., 'Measurement of local entrainment rate in the initial region of axisymmetric turbulent air jets' J. Fluid Mech, 51, 4, 773 (1972).
7. Smith, J.M. and Van Ness, H.C., *'Introduction to Chemical Engineering Thermodynamics'* 3rd edition, McGraw Hill (1975), USA.
8. Sirkar, A., 'A Kinetic model for solid phase hydrogenation of coal', ICE, 45, 3, 178 (2003).
9. Gupta, V. and Srinivasan, J., *'Heat and Mass Transfer',* Tata McGraw Hill (1978), New Delhi.
10. Indian Patent 193508 (2004) by Dr. Amalesh Sirkar.
11. Indian Patent 200006 (2006) by Dr. Amalesh Sirkar.
12. Rase, F. Howard, *'Chemical Reactor Design for Process Plants'* Vol. 2, John Wiley and Sons Inc. 1977.
13. Indian Patent 235876 (2009) by Dr. Amalesh Sirkar.

Section II

CHAPTER 3

A Concept on Gas-Gas Mixing Equipment

Gas-Gas mixing equipment is needed for various purposes as follows:

1. Homogenization of feed gases in front of catalytic reactors
2. Mixing of Claus gas with air in a sulfur recovery plant
3. Calorific value adjustment of natural gas by admixing with butane or propane. These are a few applications amongst many more in chemical industry. The common goal of all these applications is the need for a homogenous mixture within a section of duct as short as possible. Justification of the need for homogenization is as follows:

For instance, in a catalytic reactor if the gases are not thoroughly mixed there will be non-homogenity in the mixture, resulting in different rates of reactions at different points inside the reactor. In such a case one cannot predict and control the yield of product downstream of the reactor.

3.1 EXISTING MIXING DEVICE

It is often decided to use a pipeline mixer for mixing of gases. But, it is not in the knowledge of many that a length as high as 1000 pipe diameters is needed for complete mixing of two gases of density difference to the tune of 10% or so(1). The only suitable mixer available at present is Sulzer gas Mixer(2) which is a static mixer having mixing element inside a duct. A typical mixer of sulzer type(2) requires a length of 1.2 meter for a gas flow of 48,000 Nm^3/hr in a duct of cross-section 1.7 m^2. However, the mixing achieved with this mixer is not 100%. Thus, there is a need for even shorter duct than sulzer mixer and also achieve 100% mixing. Such a device is described below.

3.2 DESCRIPTION OF THE MIXER

It consists of a short horizontal duct of square cross-section containing inert particles of particular shape (shown below). The gases enter at one end, one above the other (Fig. 3.5). The duct is half filled up with inert particles. The reason for this is elaborated in point 3 under 'Effect of Various Parameters on Mixing'. The duct containing inert particles is vibrated in vertical direction by means of crank slider arrangement which hits the duct at bottom intermittently. The duct is freely supported at both ends. When the slider hits the pipe the particles will be lifted upwards. The particles after hitting the top wall of the duct will fall vertically down to the bottom wall and will be lifted again when the slider returns, and this continues as the gases enter continuously and leave the duct after mixing.

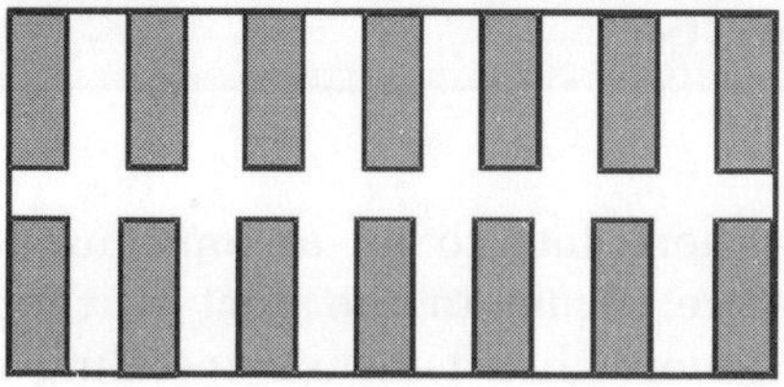

Figure 3.1: Particle shape

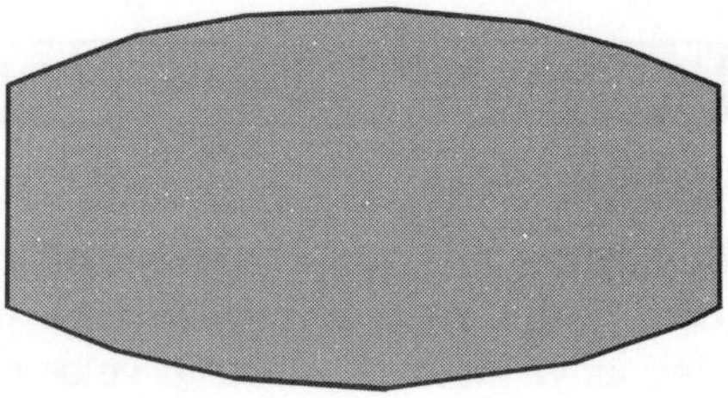

Figure 3.2: Strip cross-section

Figure 3.3: Shape of strip

3.3 MECHANISM OF MIXING

When the particle moves upward, due to compressive action on the gas medium waves are generated by the strips both along the length and width of it. The waves will travel vertically below while in the return trip of the particles vertically down, waves will be generated in the upward direction. The terminal velocity of the particles has to be greater than the upward velocity such that the waves generated from opposite directions collide resulting in mixing of the gases along the cross-section of the duct as the gases travel down the length of the duct. Actually, after collision the waves penetrate one another due to flimsy surface of the waves. The upward and downward movement within the waves occurs due to shearing action through layers of the gaseous medium due to low viscosity and their momentum, and mixing occurs due to the combination of upward and downward movements, a circulatory motion indeed within the waves. The upward and downward movement is in a curved path due to the fact that when the fluid moves vertically downward or upward, its motion is resisted by the gaseous medium encountered in its way from the opposite direction such that a lateral velocity component is generated due to encounter, resulting into the curved path and hence circulation.

The reasons for the type of curvature of the strips shown in Figs. 3.2 and 3.3 are as follows:

If the surface is flat then waves, while moving downward through the gap (between the strips) will partially collide thus, generating mixing laterally to some extent unless the gap is wide. Lateral mixing is not wanted because it creates back mixing instead of mixing in vertical direction since the gases enter the duct one above the other. If the gap between the strips is wide to avoid lateral mixing then we will have less number of strips per particle thus generating less number of waves causing less mixing and hence longer length of duct.

Having the curvature of the strips as shown the waves will be guided by the curvature to move downwards. The gap should be optimum such that the waves while moving downwards do not collide laterally and are only encountered by waves moving in opposite direction (vertically).

3.4 EFFECT OF VARIOUS PARAMETERS ON MIXING

1. Particle Size

Regarding the length and width of the particles there has to be an optimum. For instance, if the particle is too large one has to give more thrust to the duct so that the particles move with reasonable velocity upwards. Giving more thrust means spending more energy for same degree of mixing in a given length. On the other hand, if the particles are too small then its terminal velocity will be lower and hence particles won't fall rapidly such that there won't be enough interaction i.e. collision between some of the waves will take place only.

2. Number of Particles

Number of particles should be optimum. For instance, if the particles are few in number then substantial amount of space will not undergo wave generation and remain idle as far as mixing is concerned. Whereas, if the particles are large in number then there will occur hindrance between the particles which will not only slow down their velocity during free fall but also it will prevent free movement of the particles for generation of waves.

3. Duct Cross-section and Bulk Density of Particles

The duct has to be of square cross-section such that particles move same distance vertically and hence uniform mixing along the cross-section, unlike circular cross-section where particles move different distance.

The pipe should be preferably half filled up such that in one stroke all the particles move up to the upper wall with little time gap between them and maximum numbers of waves are generated. If on the other hand the pipe is ¾ th or ¼ th filled then same problem will occur as explained under Number of Particles because of either large number of particles or small number of particles. It is shown below mathematically that half filled up duct generates maximum number of waves (see Figure 3.4).

N_c – number of particles per unit volume

$$N_c\ B^2 dL = \frac{\rho_b}{w_p} V = \frac{\rho_b}{w_p} B(B - \Delta y) dL$$

$$\therefore \quad N_c = \frac{\rho_b}{w_p}\left(1 - \frac{\Delta y}{B}\right)$$

Displacement per unit time by the particles per unit volume

$$= \frac{\Delta y}{\Delta \theta} \times N_c = F = \frac{\rho_b}{w_p}\frac{\Delta y}{\Delta \theta}\left(1 - \frac{\Delta y}{B}\right)$$

For maximum F $\longrightarrow$ $\frac{dF}{d(\Delta y)} = 0 = \frac{1}{\Delta \theta}\left(1 - \frac{\Delta y}{B}\right) + \frac{\Delta y}{\Delta \theta} \times -\frac{1}{B}$

$$\therefore \quad \frac{\Delta y}{B} = 1 - \frac{\Delta y}{B} \qquad \frac{2\Delta y}{B} = 1 \qquad \therefore \ \Delta y = \frac{B}{2}$$

$$\therefore \quad N_c = \frac{\rho_b}{2w_p}$$

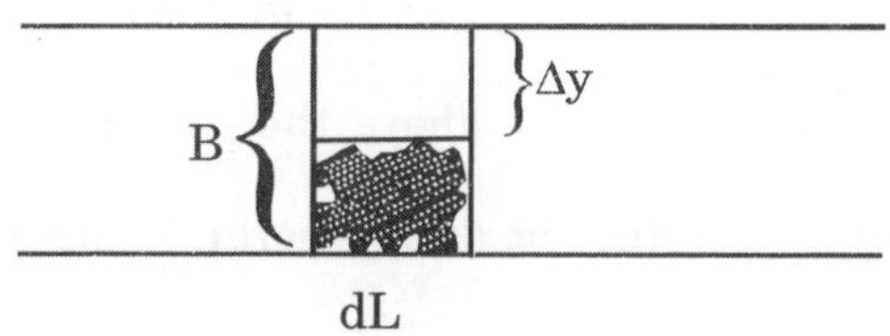

Figure 3.4

4. Number of Strips per Particle

Number of strips should be large such that for a given length of the particle, the width of the strips and space between two strips is small. This will generate maximum number of waves per particle. If the strips are wider then the space between the strips will also be wider for having to have larger clearance between the waves for non-interference between the waves generated sidewise and hence effective number of waves will be less. Although area of each strip will be less if strips are smaller, total number of waves multiplied by the area of each strip will be larger unlike the case with wider strip and wider clearance.

5. Flow Condition

In order to keep pressure drop at a minimum due to the obstruction (drag) created by the particles to gas flow in forward direction linear gas velocity should be kept preferably low. However, it is likely that there will exist an optimum velocity of gas based on total cost minimization; the cost elements being cost of power and material cost which comprises of cost of total material of the duct and cost of total number of particles. As velocity increases pressure drop increases but duct becomes of smaller cross-section and number of particles are less and vice versa.

3.5 GOVERNING EQUATIONS FOR THE MIXING PROCESS

1. Momentum Balance (see Fig. 3.5)

During Rise and Fall of Particles the following equations hold good:

$$V_v \times N_c \times B\ dy \times dz \times W_p = V_w \times y_f \times A_p \times \rho^y_{gas} \times N_c \times B \times dy \times dz \qquad ...(1)$$

$$V_t \times N_c \times B \times dy \times dz \times W_p = V'_w \times y_f \times A_p \times \rho^y_{gas} \times N_c \times B \times dy \times dz \qquad ...(2)$$

where ρ^y_{gas} = density of the gas mixture at y

Note: ρ^{y}_{gas} varies in the y direction

$$-\frac{dC^{I}_{A_y} \times B \times dy \times dz}{d\theta} = V_w \times A_p \times N_c \times B \times dy \times dz \times \frac{C^{I}_{A_y} + C^{I}_{B_y}}{2}$$

$$+ V'_w \times A_p \times N_c \times B \times dy \times dz \times \frac{\left(C^{I}_{A_{y-dy}} + C^{I}_{B_{y-dy}}\right)}{2}$$

The above equation is valid for the region above the centerline of the duct where $C^{I}_{A_y}$ decreases with time

$$= \frac{V_v \times W_p \times N_c \times B \times dy \times dz}{\rho^{y}{}_{gas_I}\, y_f} \times \frac{\left(C^{I}_{A_y} + C^{I}_{By}\right)}{2} + \frac{V_t \times W_p \times N_c \times B \times dy \times dz}{\rho^{y-dy}{}_{gas_I}\, y_f} \times \frac{\left(C^{I}_{y-dy} + C^{I}_{y-dy}\right)}{2} \quad ...(3)$$

where $\rho^{y}_{gas_I}$ & $\rho^{y-dy}_{gas_I}$ are gas mixture density above the centerline at y and $y - dy$

For the region below the centerline of duct where $C^{II}_{A_y}$ increases with time

$$+\frac{d\, C^{II}_{A_y} \times B \times dy \times dz}{d\theta} = \frac{V_v \times W_p \times N_c \times B \times dy \times dz}{\rho^{y}_{gas_{II}}\, y_f} \times \frac{\left(C^{II}_{A_y} + C^{II}_{B_y}\right)}{2}$$

$$+ \frac{V_t \times W_p \times N_c \times B \times dy \times dz}{\rho^{y-dy}_{gas_{II}}\, y_f} \times \frac{\left(C^{II}_{A_{y-dy}} + C^{II}_{B_{y-dy}}\right)}{2} \quad ...(4)$$

where $\rho^{y}_{gas_{II}}$ & $\rho^{y-dy}_{gas_{II}}$ are gas mixture density below centerline at y and $y - dy$. At the centerline $C^{I}_{A_y} = C^{II}_{A_y}$

and $C^{I}_{B_y} = C^{II}_{B_y}$

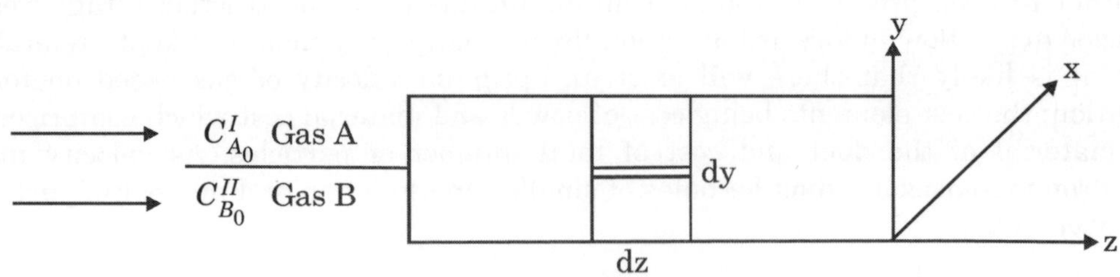

Figure 3.5

$$d\theta = \frac{dz}{V_z(y)}$$

$\therefore$ Equation (3) becomes

$$\frac{-V_z(y)dC^{I}_{Ay}}{dz} = \frac{V_v \times W_p \times N_c}{y_f}\left[\frac{C^{I}_{A_y} + C^{I}_{B_y}}{2\rho^{y}_{gas_I}}\right] + V_t \times \frac{W_p \times N_c}{y_f}\left[\frac{C^{I}_{A_{y-dy}} + C^{I}_{B_{y-dy}}}{2\rho^{y-dy}_{gas_I}}\right]$$

For simplicity of analysis, variation between the concentrations of A and B at y and $y - dy$ may be neglected. Similarly, variation between $\rho^{y}_{gas_I}$ and $\rho^{y-dy}_{gas_I}$ may be neglected too.

Thus, $$\frac{-V_z(y)dC^I_{Ay}}{dz} \simeq \frac{(V_v + V_t)\times W_p \times N_c}{y_f}\left[\frac{C^I_{A_y} + C^I_{B_y}}{\rho^y_{gas_I}}\right] \quad ...(5)$$

Similarly equation (4) becomes $$\frac{V_z(y)\,dC^{II}_{A_y}}{dz} = \frac{(V_t + V_v)\times W_p \times N_c}{y_f}\left[\frac{C^{II}_{A_y} + C^{II}_{B_y}}{\rho^y_{gas_{II}}}\right] \quad ...(6)$$

2. Mass Balance for A and B

$$V_{z_o} \times \frac{B}{2}\times B \times C^I_{A_o} = \int_{B/2}^{B} V_z(y)\times C^I_{A_y} \times B \times dy + \int_{0}^{B/2} V_z(y)C^{II}_{A_y} \times B \times dy \quad ...(7)$$

Also, $$V_{z_o} \times \frac{B}{2}\times B \times C^{II}_{B_o} = \int_{0}^{B/2} V_z(y)\times C^{II}_{B_y} \times B \times dy + \int_{B/2}^{B} V_z(y)C^I_{B_y} \times B \times dy \quad ...(8)$$

Furthermore, $$-V_z(y)\frac{dC^I_{A_y}}{dz} = V_z(y)\frac{dC^I_{B_y}}{dz}$$

$$V_z(y)\frac{dC^{II}_{A_y}}{dz} = -V_z(y)\frac{dC^{II}_{B_y}}{dz}$$

These equalities hold good because, R.H.S. of equations 5 and 6 are same for increase in concentration of B above centerline of duct and decrease in concentration of B below centerline of duct respectively (just the opposite of concentration of A), the mechanism of mixing being same for A and B across the cross-section. Note that B increases above centerline and decreases below centerline.

$$\therefore\ -\frac{dC^I_{A_y}}{dz} = \frac{dC^I_{B_y}}{dz} \quad ...(9)$$

and $$\frac{dC^{II}_{A_y}}{dz} = -\frac{dC^{II}_{B_y}}{dz} \quad ...(10)$$

In addition, $$\rho^y_{gas_I} = C^I_{A_y}M_A + C^I_{B_y}M_B$$

$$\rho^y_{gas_{II}} = C^{II}_{A_y}M_A + C^{II}_{B_y}M_B$$

Determination of V_t and particle size

At a distance $y - dy$ the following equality holds good:

$$\frac{dy}{V_w} = \frac{B - y + 2dy}{V_t} + \frac{dy}{V'_w} \quad ...(11)$$

Time taken by wave to travel '*dy*' downwards at y is equal to time taken by particle to fall '*B* – *y* + 2*dy*' plus travel time for '*dy*' upwards by the wave generated in upward direction. Hence, the collision at $y - dy$.

From equations (1) and (2)

$$V_w = \frac{V_v W_p}{y_f A_p \rho_{gas}^{y-dy}} \quad \text{and} \quad V'_w = \frac{V_t W_p}{y_f A_p \rho_{gas}^{y-dy}}$$

$$\therefore \quad \frac{y_f A_p \rho_{gas}^{y-dy} dy}{V_v W_p} = \frac{B - y + 2dy}{V_t} + \frac{y_f A_p \rho_{gas}^{y-dy} dy}{V_t W_p}$$

$$\frac{V_t}{V_v} = 1 + \frac{B - y + 2dy}{V_t} \times \frac{V_t W_p}{y_f A_p \rho_{gas}^{y-dy} dy}$$

$$\simeq 1 + \frac{B - y + 2\Delta y}{\Delta y} \times \frac{W_p}{y_f A_p \rho_{gas}^{y-dy}} \qquad \text{...(12)}$$

For V_t to be maximum y has to be minimum and hence

$$\therefore \quad \frac{V_t}{V_v} = 1 + \frac{B + 2\Delta y}{\Delta y} \frac{W_p}{y_f A_p \rho_{gas}^{y-dy}}$$

Taking a small value of Δy in comparison to the dimension of the duct (one side of duct *B*) one can get an expression for V_t in terms of V_v and other parameters. Depending on V_v one can Calculate V_t (maximum) taking a minimum value of ρ_{gas}^{y-dy}. Knowing V_t one can get mean particle size from terminal velocity correlation.

3.6 VERIFICATION OF MODEL EQUATIONS

Experiments can be carried out on mixing of any two gases *A* and *B* and $C^I_{A_y}, C^{II}_{A_y}, C^I_{B_y}$ and $C^{II}_{B_y}$ should be measured along the cross section and along the length. Also, V_z (y), the velocity distribution in y direction is to be developed from measurements in the dust using pitot tube. Next, length of the mixer used for experiment should be checked against the theoretical equation (5), (6), and (10) as follows:

From equations (5) and (6) after substituting for N_c which is equal to $\rho_b/2W_p$ and writing in difference form one gets

$$\Delta z = -\frac{2V_z(y) \times \Delta C^I_{A_y} \times y_f \times \rho^y_{gas_I}}{(V_t + V_v) \times \rho_b \times \left\{ \dfrac{(C^I_{A_y} + C^I_{B_y})|_z + (C^I_{A_y} + C^I_{B_y})|_{z+\Delta z}}{2} \right\}} \qquad \text{...(13)}$$

Similarly, $$\Delta z = -\frac{2V_z(y) \times \Delta C^{II}_{A_y} \times y_f \times \rho^y_{gas_{II}}}{(V_t + V_v) \times \rho_b \times \left\{ \dfrac{(C^{II}_{A_y} + C^{II}_{B_y})|_z + (C^{II}_{A_y} + C^{II}_{B_y})|_{z+\Delta z}}{2} \right\}} \qquad \text{...(14)}$$

Similar equations can be written in terms of $C^{I}_{B_y}$ and $C^{II}_{B_y}$ using equations (9) and (10).

Equations (13) and (14) can be used to calculate incremental lengths Δz for particular 'y's, summation of which would give total length of the duct 'L' required to achieve the following and the 'L' obtained should be equal to the experimental length.

$$\frac{C^{I}_{Ao} + C^{II}_{Bo}}{2} = C^{I}_{A_y}\Big|_L + C^{I}_{B_y}\Big|_L = C^{II}_{B_y}\Big|_L + C^{II}_{Ay}\Big|_L \qquad ...(15)$$

Value of 'y_f', the thickness of wave produced is to be assumed for estimation of length of the mixer by either equation (13) or (14) and should be checked against experimental length. It may be noted that length of the duct chosen for experiment may be greater than the minimum length needed for 100% mixing, which will be evident from the experimental data showing no change in composition beyond the minimum length. On the other hand if the length chosen is less than that for 100% mixing then also experimental and theoretical lengths could be checked using either equation (13) or (14).

'y_f' varies with variation in 'V_v' and 'V_t' and 'W_p' (weight /particle) because the compressive action of the particles on gases for formation of waves is proportional to the thrust by the moving particle i.e. (velocity × weight). Therefore, the following correlation for y_f may be developed.

$$y_f = A\,(V_v \times W_p)^{\alpha} \times (V_t \times W_p)^{\beta} \qquad ...(16)$$

For design of a mixer for mixing of any two gases A and B, experimental data are needed for a range of initial concentration $C^{I}_{A_o}$ and $C^{II}_{B_o}$ and for a range of initial velocity (for a particular flow regime) and family of curves are to be developed based on measurement of concentrations along the cross-section and along the length such that for any combination of $C^{I}_{A_o}$ and $C^{II}_{B_o}$ and initial velocities, length could be determined. However, correlations could be developed from the family of curves for direct determination of length. Note that duct dimension (B) and initial velocity could be obtained from the flow rate and choosing the Reynolds No.

Some experiments have been carried out and it has been found that 100% mixing can be achieved and the corresponding length of the mixer is shorter than Sulzer Mixer.

NOMENCLATURE

A_p – total surface area of the strips (one side only) per particle

B – width and height of the square duct

$C^{I}_{A_o}$ – concentration of gas A at the entrance to the mixer

$C^{I}_{A_y}$ – concentration of gas A at any point in the mixer above the centerline of the duct

$C^{I}_{A_{y-dy}}$ – concentration of gas A at any point in the mixer above the centerline of the duct, just below $C^{I}_{A_y}$

$C^{II}_{A_y}$ – concentration of gas A at any point in the mixer below the centerline of the duct

$C^{II}_{A_{y-dy}}$ – concentration of gas A at any point in the mixer below the centerline of the duct, just below $C^{II}_{A_y}$

$C^{I}_{B_y}$ – concentration of gas B at any point in the mixer above the centerline of the duct

$C^{I}_{B_{y-dy}}$ – concentration of gas B at any point in the mixer above the centerline of the duct, just below $C^{I}_{B_y}$

$C^{II}_{B_y}$ – concentration of gas B at any point in the mixer below the centerline of the duct

$C^{II}_{B_{y-dy}}$ – concentration of gas B at any point in the mixer below the centerline of the duct, just below $C^{II}_{B_y}$

$C^{II}_{B_0}$ – concentration of gas B at the entrance to the mixer

F – displacement per unit time by the particle per unit volume

L – length of the mixer

M_A, M_B – molecular weight of gas A and B respectively

N_c – number of particles per unit volume of the mixer

V_v – lifting velocity of the particles

V_t – terminal velocity of the particles

V_z (y) – linear velocity of the gas (a function of y direction)

V_{z_o} – velocity with which gas A and B enters the duct

V_w – velocity of wave in the downward direction

V'_w – velocity of wave in the upward direction

W_p – weight per particle

y_f – thickness of wave

Δy – displacement in y direction

REFERENCES

1. Hiby, J.W., Verfahrenstechnik 4 (12) 538 (1970).
2. Tasurcher, W.A. and Streiff, F.A., 'Static Mixing of Gases' CEP April, 61 to 65 (1979).

Section II

CHAPTER 4

Pollution Control Equipment Design

Effluents of Organic Chemical Industries which are biologically degradable are usually treated in various types of air-water contacting devices e.g. Trickling Filter, Underground Aeration tank, Rotating Biological Contactor etc. However, apart from air-water contact devices the biological sludge that is produced needs to be separated from the treated water before its discharge to surface water. For this purpose we normally use a separate secondary clarifier for sludge removal which can be either a mechanical clarifier or a vertical tube clarifier. This is required in addition to primary clarifier.

These days our objective is to use minimum amount of land for installation of effluent treatment plants because land is costly and also because often there is no payback for investment on effluent treatment plants. While for new plants this is the fact, for old plants this is more so because when such plants were built there was no provision for ground space for building ETP because pollution control as a requirement was not in force in those days. Thus, today there is a need for primary and secondary treatment processes, which would require minimum ground space and yet fulfill the objective of desired BOD removal along with sludge removal. How can we achieve these goals? The logical thing to do is to design a single unit that would do both air-water contact and sludge separation by taking advantage of a housing, which is tall and narrow in diameter. Thus, underground aeration tank becomes a back dated technology because we cannot make it deeper than 4 meter in order to keep the content in mixed condition and we need additional ground space for both primary and secondary clarifier for sludge removal. Rotating Biological Contactor (RBC) is not the answer either, because the basic principle of design requires large number of discs mounted horizontally on a shaft requiring ground space and the primary and secondary clarifier have to be constructed separately. So, we are left with the only device i.e. trickling filter amongst the conventional devices. Alternatively, we can construct an above ground aeration tank that is tall and narrow and to house the primary and secondary clarifiers we can make annular vessels around the central tank with downcomers that would do the job of solid separation in a fashion similar to a vertical tube clarifier. For air-water contact we can have a air sparger at the bottom of the central tank and the desired hydraulic retention time can be given to the waste water by allowing it to flow vertically upwards. Detail design of a typical unit of this type is given below. The above mentioned concept of housing both primary and secondary clarifier is also applicable to modern designs of trickling filters where PVC fill packs are stacked vertically for air-water contact in a tall tank. Details are given later.

4.1 VERTICAL AERATION TANK COUPLED WITH PRIMARY AND SECONDARY CLARIFIER

Figures 4.1 and 4.2 show typical design with dimensions for vertical aeration tank with an annular vessel in the top region with downcomers meant for primary settling tank and a similar design near the bottom region meant for secondary settling tank. This particular design has been used for effluent treatment of a vanaspati factory waste water(1). The downcomers are attached to the vessel wall by fasteners. The effluent enters the primary clarifier downcomers from water boxes (connected via a header) which help in destroying incoming flow disturbances by means of enlargement of cross-sectional area for flow. The downcomers are tapered (Figure 4.2) for slowing down the velocity of effluent water so that the solids begin to separate out before it enters the annulus.

As the effluent water comes out of the downcomer, its velocity decreases due to enlarged cross-section of the annulus. The velocity to be maintained in the downcomers is about 31 meter/hr and the corresponding Reynolds No. should be between 800 and 900, which is optimum in terms of separation and material savings. For instance, if Reynolds No. is lower than 800, separation of solids will be marginally better but the number of downcomers will be more and hence material requirement will be more.

The sudden enlargement of cross section after passing through the downcomers causes lowering of velocity, which helps in separation of solids as the effluent rises slowly upwards through the annulus. The solids move towards the conical bottom region of the clarifiers being heavier than water and the water moves upward because the opening of conical bottom is small, giving more resistance to flow downward than upward. The cross-sectional area of the annulus should be such that the velocity of the rising wastewater is about 0.5 m/hr and the corresponding Reynolds No. should be around 125. At this Reynolds No. for upward flow through the annulus, the separation of solids has been found to be most effective. The solids come out of the conical bottom opening in the form of dilute slurry as in a conventional clarifier.

The height of the central vertical aeration tank is fixed by the heights of the clarifiers and the diameter of the tank depends on the hydraulic retention time.

The pipe from the launder of primary clarifier comes down and is connected to the bottom of the aeration tank through which water enters the central tank where it is contacted with air. The air is blown by a compressor and enters via a sparger located at the bottom of the central tank (Figure 4.1). The quantity of air and the sparger design is such that the contents of the tank are in mixed condition. The sparger is in the form of concentric rings with holes in the rings positioned downwards so that air pushes the liquid from the bottom most section of the aeration tank.

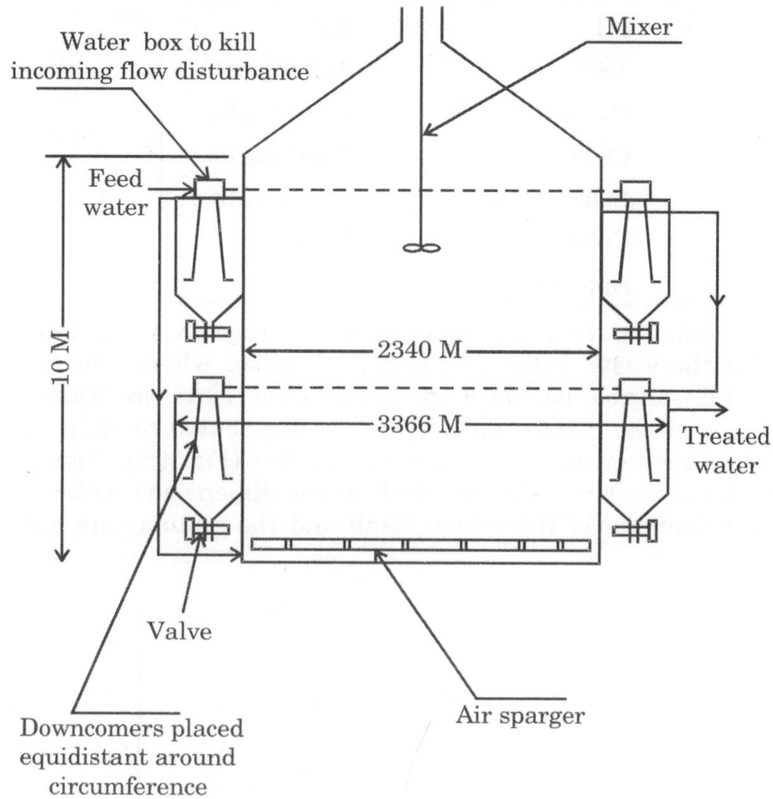

Figure 4.1: 3 Tier unit

There is an opening at the top of the 3 tier unit through which oxygen-depleted air exits. The waste water which flows upward in the aeration tank, exits through a pipe located slightly below the primary clarifier launder height. The magnitude of this difference in height depends on the frictional drop in the pipeline from the launder of primary clarifier to the base of aeration tank and the expansion loss at the point where waste water enters the aeration tank. The pipe through which the waste water comes down from the top of the aeration tank, enters the header of the secondary clarifier leading to the downcomers similar to the primary clarifier design.

A portion of the sludge from secondary clarifier bottoms is pumped and it enters at the bottom of the aeration tank, the process of biological degradation being an activated sludge process. It may be noted that although the configuration of aeration tank and clarifiers is different from the conventional design, the biological process parameters are the same as in a conventional design. Nutrients are added via dosing pumps to the bottom of the central vertical tank.

4.2 AN EXAMPLE OF THE NOVEL DESIGN

The above mentioned design has been incorporated in an old vanaspati manufacturing company for the acid oil effluent generated in acid-oil section. The name of the company is Vegetable Products Ltd.

The analysis of the effluent (2 m^3/hr flow rate) treated is as follows:

pH	2.0
TSS	712 mg/lit
Oil and Grease	1700 mg/lit
COD	5000 mg/lit
BOD	2200 mg/lit
Colour	Yellow
Appearance	Turbid

The effluent is pumped from an underground storage to an oil-water-solid separator. From the separator the waste water goes to a flash mixer where lime is added to increase the pH to around 7.5, suitable for biological degradation. Next, the waste water with solids is pumped to the top of the 3 tier unit where it is distributed to eight downcomers placed equidistant from one another inside the annular clarifier (Fig. 4.2). Figure 4.2 shows details of the schematic drawing of the clarifier with major dimensions. Other major dimensions such as height and diameter of the central tank and the annulas are given in Figure 4.1.

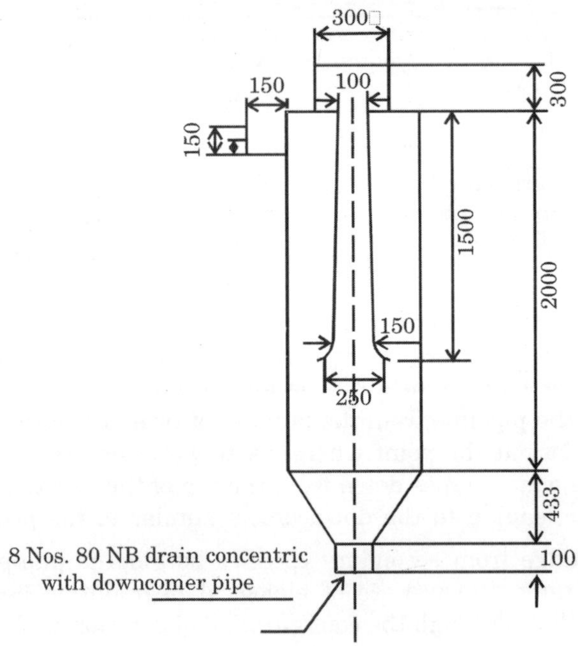

Details of clarifier dimensions

(All dimensions are in mm)

Figure 4.2

The hydraulic retention time in the aeration tank is 20 hours which is based on the BOD level in the incoming effluent (2200 mg/l) and the treated water (30 mg/l). The MLSS maintained in the aeration tank is 3500 mg/l. Food to microorganism ratio maintained is about 0.3.

The air requirement for contents of the tank to be in mixed condition is 46 scfm. Air enters through four concentric rings containing 150, 90, 50 and 10 holes respectively for the outer to the innermost ring. The diameters of the rings are 2 meter, 1.5 meter, 1 meter and

0.5 meter respectively. The holes are 1 mm dia placed equidistant from one another and the velocity through a hole is 300 fps. Apart from this air quantity used for keeping the contents of the central tank in mixed condition, a mixer (vertical) is located near the top region for creating additional mixing between air and water because above a certain height simple sparger for air flow is not enough to keep the contents of the tank in mixed condition.

The treated water characteristics are as follows:

pH	7 to 8
TSS	Less than 100 mg/lit.
Oil and Grease	8 mg/lit.
COD	Less than 250 mg/lit.
BOD	Less than 30 mg/lit.
Colour	No colour
Appearance	Clear water

4.3 TRICKLING FILTER

Modern trickling filters are of square cross-section, packed with PVC fill packs of rectangular shape. In the modern trickling filters waste water flows through the channels created by the PVC sheets in fill packs in the form of a film and air flows counter current to the waste water flow i.e. upwards. The fill packs rest on pipes installed just below the packs. In the upper portion of the trickling filter we can create an annular space (Fig. 4.3) with downcomers for separating primary sludge.

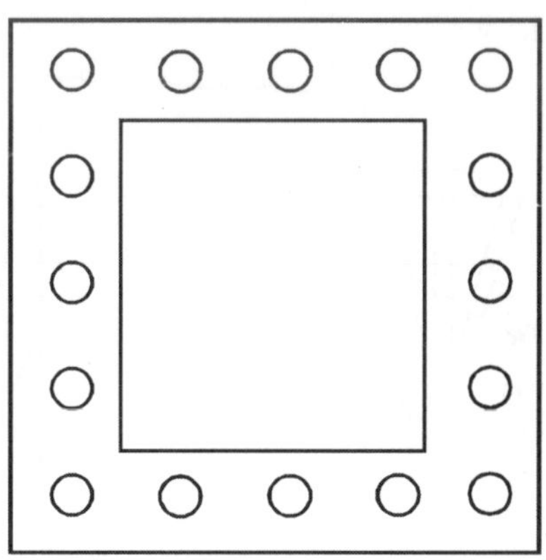

Figure 4.3

Unlike the design described earlier, here waste water after separation of primary sludge flows down to the pump suction (Fig. 4.4) and is pumped to the top where it enters nozzles for uniform distribution along the cross-section of the filter. The waste water containing biological sludge, after passing through the channels (formed by PVC sheets in a pack) is pumped again to the top of the lower half of trickling filter where secondary clarifier is located in a square annulus containing downcomers. To keep the secondary sludge under suspension in treated water before it is pumped, stirrers can be installed sidewise at the bottom of the tower; the trickling filter.

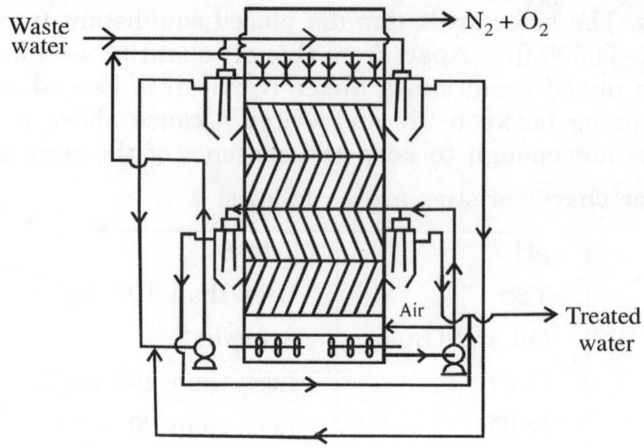

Figure 4.4

For minimizing the water content of the primary and secondary sludge one can put restriction at the bottom of the conical annular vessels. Because this will ensure that most of the water flows upward due to resistance to flow at bottom and solids being heavier will pass through the restriction.

Furthermore, greater separation of solids, by slowing down the velocity of water upwards can be achieved by tapering the annular vessel and installing baffles as shown below (Figure 4.5). This is because drag on the particles in order to entrain them upwards will be reduced due to slow down caused by reduction in liquid velocity. The particles which will be in the vicinity of the baffle will easily fall downwards due to their weight because liquid velocity is reduced to zero at the baffle walls.

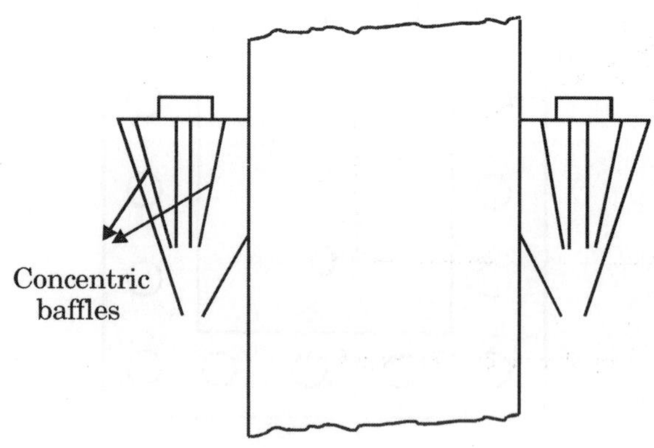

Figure 4.5

REFERENCE

1. 'Space Saving Technology in Waste Water Treatment' by A. Sirkar, Indian Chemical Engineer, Vol. 43, No. 1, 2001.

Index